AFRICA'S
MOUNTAINS
OF THE
MOON

Thoughts on the re-discovery of Africa's Mountains of the Moon

"Another emotion is that inspired by the thought that in one of the darkest corners of the earth, shrouded by perpetual mist, brooding under the eternal storm-clouds, surrounded by darkness and mystery, there has been hidden to this day a giant among mountains, the melting snow of whose tops has been for some fifty centuries most vital to the peoples of Egypt."

Henry Morton Stanley
In Darkest Africa, Vol. II, p. 299 (1890)

Canarina eminii

AFRICA'S MOUNTAINS

OF THE

MOON

JOURNEYS
TO THE SNOWY SOURCES OF THE NILE

Text & Photographs
GUY YEOMAN

Botanical Illustrations
CHRISTABEL KING

Elm Tree Books · London

ELM TREE BOOKS
Published by the Penguin Group
27 Wrights Lane, London W8 5TZ, England
Viking Penguin Inc, 40 West 23rd Street, New York, New York 10010, U.S.A.
Penguin Books Australia Ltd, Ringwood, Victoria, Australia
Penguin Books Canada Ltd, 2801 John Street, Markham, Ontario, Canada L3R 1B4
Penguin Books (N.Z.) Ltd, 182-190 Wairau Road, Auckland 10, New Zealand

Penguin Books Ltd, Registered Offices: Harmondsworth, Middlesex, England

First published in Great Britain 1989 by Elm Tree Books

1   3   5   7   9   10   8   6   4   2

This edition and design copyright © Savitri Books Ltd, 1989
Text and photographs copyright © Guy Yeoman, 1989
Botanical illustrations copyright © Christabel King, 1989

Produced and designed by
SAVITRI BOOKS LTD
Southbank House
Suite 106, Black Prince Road
London SE1 7SJ

Art direction and design by Mrinalini Srivastava
Edited by Wendy Lee

Typeset in Palatino by Dorchester Typesetting Group Ltd, Dorchester, Dorset
Reproduction by Mandarin Offset Ltd, Hong Kong
Printed and bound in Spain by Grupo Nerecan, Madrid

British Library Cataloguing in Publication Data
CIP data for this book is available from the British Library

ISBN 0-241-12806-4

Frontispiece. Canarina eminii. *Emin's Bell flower is one of the most elegant of the Rwenzori flora. It is an epiphitic vine, and the flowers hang in clusters like Christmas lanterns from riverine forest trees. Christabel King made this picture at 8,400 feet at the Mubuku crossing at the start of our journey up the Bujuku valley.*

To my wife, for love and courage

Author's Note

Anyone writing about the Rwenzori mountains and the region of the Mountains of the Moon has to face questions about place names and their spelling. The names that were bestowed by the early explorers, even when given the approval of government survey departments or of the Royal Geographical Society, were often inappropriate; they frequently ignored local custom and used spellings that have since been changed by independent African countries. The area covered by this book is particularly difficult – not only does it embrace several African languages, but it is divided between anglophone Uganda and francophone Rwanda and Zaire. In as much as it is possible to be consistent, I have tried to follow the modern usage of these countries, except when directly quoting from earlier writers, when I have left their spelling unchanged.

An immediate source of confusion may be the conflicting use of Rwenzori and Ruwenzori. The latter was Stanley's choice, but he did not have good grounds for the name itself. Harry Johnston, in his *Uganda Protectorate* (1902) points out its lack of vernacular authenticity and lists a number of contenders. None the less, the pleasing quality of euphony and nobility of the name *Ruwenzori* had led all subsequent European writers to use it and it has been perpetuated in the specific names of numbers of plants and animals. In modern-day Uganda, the name is accepted but with the subtly different spelling – *Rwenzori*, and I have therefore followed this custom. (However, a purist might object that the preferred name amongst the Konjo themselves is *Rwenzururu*, and this may yet come into more general use.)

Similarly, in the days of German, and then Belgian, rule in the country now known as Rwanda, *Ruanda* used to be the standard spelling. Other examples of such changes are too numerous to list and, while I have done my best in what is a rather fluid field, readers should not be surprised if they encounter variations of spelling in other texts and maps. I should perhaps mention that while *Konjo*, for the Bakonjo people who are the heroes of my story, has so far stood the test of time, in the still rather rare instances where they write about themselves I have noticed that they prefer *Konzo*! There is a valuable note on Rwenzori and Rukonjo names in Osmaston and Pasteur's 'Guide' (see Bibliography).

All the African words in my text, whether of ordinary vocabulary or the names of people and places, should be pronounced according to the general rule that, where they are polysyllabic, the emphasis should be on the penultimate syllable. If I add that the letter 'e' should generally be pronounced 'ey' rather than 'ee', then I believe the English speaker will not go far wrong.

CONTENTS

Opposite. *The heart of Rwenzori: Mount Stanley above Lake Bujuku. This rare clear view shows Mount Margherita (right, 16,763 feet); Mount Alexandra (left, 16,703 feet), and set back between them, Point Albert (16,690 feet). The Margherita glacier between the peaks joins (left) with the edge of the Stanley ice field. Below, steep slopes descending to Lake Bujuku are covered with giant groundsel forest and blue-flowering Wollaston's lobelias.*

INTRODUCTION

By late afternoon our small wood-burning steamer was half-way across the inland sea of Lake Albert, a 25-mile-wide torrid trench in Africa's western Rift valley. I was returning from a minor intelligence assignment in the Congo – the sort of gift that comes to you sometimes in war. Earlier in the day we had weighed anchor at the Belgian port of Kasenyi – just a jetty, little more – and I was standing at the ship's rail, gazing at the amorphous blue-grey mass of haze that constantly filled the southern horizon, tantalisingly hiding the heart of the continent.

The sun had slipped behind the western escarpment, 3,000 feet above us, instantly plunging the surface of the lake into a forbidding indigo blackness; yet at once the waters were set on fire, the crest of each wave a luminous flamingo's feather as it picked up the scarlet of the cirrus clouds overhead. As I watched, the shapeless cloud mass to the south began to structure itself into horizontal silver wafers. The highest of these curved majestically, broke and revealed, rose-tinted by the hidden sun, two snow-covered mountain peaks. I watched them for perhaps half a minute, disbelieving, before the vision faded and vanished, as though it had never been.

There was a quiet whistle from behind me, and a voice in Scots accents said, 'Phew – you don't know how privileged you are. I've crossed the lake scores of times and only seen that once before: Crophi and Mophi – *Lunae montes* – Ruwenzori – the Mountains of the Moon – just as the Ancient Greeks prophesied two thousand years ago.' The captain of this tiny vessel, the *Robert Coryndon*, black-bearded with a time-etched mahogany face, was twice my age and had originally struck me as taciturn. However, when I dined with him that evening, as our little steamer felt its way northwards up the Ugandan coast through the inky black of the night, I was impressed by his erudition.

'The Ancient Greeks in Egypt were obsessed with it, you know. Where on earth could all this water come from – the Nile, I mean – flowing so constantly out of the desert? But they either knew something or were damned good guessers. Aeschylus – he was 500 years BC – talked about Egypt nurtured by the snows. And then there was Herodotus, only a few years later. Do you know he actually made an expedition up the Nile for hundreds of miles, and he heard this story about the river rising from a lake lying between two mountains called Crophi and Mophi? But like most other people, he couldn't credit the idea of snow on the equator. And then there were stories of the great lakes – from people like Eratosthenes, you know, and Aristotle and Hipparchus. They may have been quoting each other, but surely it can't all have been sheer guesswork?'

I poured him another dram from my bottle of army ration Canadian rye whisky, an insult to his Hebridean palate that he bore with commendable stoicism: after all, it was 1943, there was a war on, and we all had to do our bit.

'So when did the news break?' I asked, 'I mean, when did they actually discover the snow mountains?'

'Well, that's a good question, laddie – a very good question – but for my money, I'm a Ptolemist. Claudius Ptolemaeus – what a grand old chappie he must have been. Not that he ever made an expedition in his life: he just sat on his arse in his library at Alexandria – a real armchair geographer. But he got all the news; and he had this sort of book of Admiralty Sailing Directions, put together years before by Marinus of Tyre – the *Periplus of the Erythrean Sea*, and in it was this story about a Greek traveller, Diogenes, who had made a journey into the hinterland from the Indian Ocean coast. He found two great lakes that were said to be the twin sources of the Nile. They can only have been Lake Victoria and our lake here,

Lake Albert. And where did these lakes get their water from? Why, from a range of snowy mountains – the *Lunae montes* – the Mountains of the Moon.'

The captain slapped his hand on the table as though he had just played a trump card.

'Well, there you are my boy. Of course if you're a philistine modernist, you'll say they were discovered by Stanley – Henry Morton Stanley, the Americanised Welsh bastard – only fifty or so years ago. Well, you pays your money and you takes your choice.'

The bottle was empty and it went over the side. Is it in Lake Albert to this day, or has it followed the river down to Egypt? Thrilled by the story I had just heard, I turned in to my bunk but could scarcely sleep for thinking of the mountain vision that had been vouchsafed me. It is as bright now, forty-five years later, as it was then in the morning of my life.

Did Stanley really discover the Mountains of the Moon in 1888, a mere fifty-five years before my vision? He certainly deserved to, and it would be mean to try to deny him the credit – kudos is the word he would have used. But it wasn't quite such a straightforward matter. Between the Ancient Greeks and the Victorians, only the Arabs – seafarers, land travellers and slavers – had expressed any opinion on the subject, but nothing more transpired than the Greeks had already told us. In the middle of the nineteenth century, German missionaries were the first modern Europeans to open the easternmost windows on to Dark Africa. In 1848 Johann Rebmann first recorded snow in the tropics when he sighted Kilimanjaro, and the next year his colleague Johann Lewis Krapf sighted Mount Kenya. But it was soon shown that neither of these spectacularly beautiful snow-capped mountains had any connection with the Nile.

Then came what must rank as the most critical of all the great expeditions in quest of the Nile source, but also the most controversial. In 1857 Richard Burton and John Hanning Speke made the first European journey through what is now called Tanzania to Lake Tanganyika, and on the return leg Speke made a solo diversion to discover the southern coast of Lake Victoria. Burton treated Speke's intuitive claim that this lake must be the source of the Nile with facetious contempt. Smarting under this scorn, Speke returned with James Grant in 1861, circuiting Lake Victoria to the west in his ultimately successful search for the definitive source of the Nile where it debouches from the north of the lake. At the most westerly point of his route, in the Karagwe district of northwest Tanzania, he came to the court of King Rumanika. While he was there, as he records in his *Source of the Nile*, 'my attention was attracted by observing in the distance some bold sky-scraping cones situated in the country of Ruanda . . .' He goes on, 'The Mfumbiro cones in Ruanda, which I believe to reach 10,000 feet, are said to be the highest of the Mountains of the Moon.'

These are the Virunga mountains, a cluster of seven or eight volcanoes which stand a hundred miles south of Rwenzori, and these mountains were thus dubbed 'Mountains of the Moon' twenty-five years prior to Stanley's discovery of Rwenzori. But Stanley was a journalist with a flare for hype, while Speke was to die tragically only a year or two after his discovery.

Soon after Stanley's last expedition of 1888, other candidates for moon mountainship were proposed. The German explorer Oscar Baumann, in 1892, chose his 'Missosi ya Mwezi und die Nilquelle' (Mountains of the Moon and Nile Source) in latitude 3° south of the Equator, at about 7,000 feet in the Kibira forest of Birundi, while Richard Kandt, another German, explorer and administrator, in 1903 chose his source further north, in the mountainous Nyungwe forest of Rwanda. Modern Burundians correctly claim the most southerly source as the Luvironza stream more than 4° south, only a few miles from Lake Tanganyika. I have visited all these sources and several more and feel inclined to

say, with my Scottish steamer captain, 'You pays your money and you takes your choice!'

The point which I think deserves at least to be laid open to discussion is that the Ancient Greeks, in speculating on the origins of the great Egyptian river, were probably doing no more than proposing that it rose from a necessarily extensive mountainous region somewhere in equatorial Africa. To them, had they known of it, the whole wonderful mountainous area that includes north-west Tanzania, Burundi, Rwanda, south-west Uganda and the

Opposite. *Morning light on Mount Luigi di Savoia to the south of the lower Kitandara lake. The peak to the left of the skyline is Sella (15,179 feet), named after the Duke of the Abruzzi's gifted photographer, Vittorio Sella.*

---

be excused a slight feeling of embarrassment at his over-writing (we are not, after all, talking about the Himalayas,) while at the same time sympathising with his eloquently expressed emotions.

Strangely, although Stanley did not realise it, this was not his first sight of the mountains, even if it was his first sight of the snows. On his earlier transcontinental expedition in 1875, while in southern Uganda, he had seen from the east 'a faint view of an enormous blue mass afar off, which we were told was the great mountain of Gambaragara'. There is no doubt that he was looking at Rwenzori, characteristically blanketed in cloud. I have seen the all-too-familiar 'faint blue mass' from the same spot. From whichever aspect you view it, the range is nearly always concealed in this way. Indeed, by the mid-1870s several Europeans had had the possibility of viewing it, but had passed on their way, not realising it was there.

These men were officers of the extending Anglo-Egyptian administration of the Sudan, before it was swept aside by the whirlwind Muslim fundamentalism of the Mahdi rebellion, which started in 1882. The founder of this administration, Sir Samuel Baker, had discovered Lake Albert in 1864, but far from recognising Rwenzori, had proposed that his lake extended right through those parts to unite with Lake Edward! The Anglo-Egyptian government reached its height, and in due course its tragic end, under General Gordon, and Lake Albert was included in its Equatoria province. Almost incredibly, tiny wood-burning steamers had been brought up the Nile as early as 1869 and launched upon the lake, and Gordon's cosmopolitan officer corps – Emin the Austrian, the Italians Gessi and Casati, and the American Colonel Mason – had explored and charted this

extreme eastern slither of Zaire, would have qualified as sources of the Nile and Mountains of the Moon.

But Stanley's hype has won the day. Any modern Rwenzorist who reads his Victorian-style eulogy for his new-found mountains may inland sea. They all reported the 'blue mass' to the south, but never the snowy peaks. The possible exception was Romolo Gessi, who had written privately that once he had seen 'a strange vision in the sky, like a mountain covered with snow'. Who can doubt that he had seen the same vision as I had?

Now, even as we approach the moment of Stanley's discovery, comes more confusion. Stanley's great expedition – in effect, an expeditionary force that had been sent by Britain to rescue Emin from the Mahdi, had been camped on the western escarpment above Lake Albert for several months without any sighting of the

---

Below. *A delectable high source of the Nile amongst moss and giant groundsels at 13,500 feet on the slopes of Mount Speke.*

mountains. On 20 April 1888, Stanley asked his medical officer, Thomas Heazle Parke, with another of his officers, A. J. Mounteney Jephson, to take their sectional boat, the *Advance*, from the high camp down to the lake shore, to be assembled and launched for the search for Emin. Parke wrote, 'On the march we distinctly saw *snow* on the top of a huge mountain situated to the south-west of our position . . . some of the Zanzibaris tried to persuade us that the white covering which decorated this mountain was *salt*, but Jephson and myself were quite satisfied that it was snow.'

Parke mentioned his sighting to Stanley two days later, when reporting the launch of the *Advance*, but Stanley in his *In Darkest Africa*, records Parke as speaking of some much lower hills in a different direction, while in his autobiography he is devious on the subject. He was not, alas, a man who could bear his subordinates to outshine him. But fate was kind: a month later, on 24 May 1888, while on the march across the grassy plain beside the lake, he writes, 'my eyes were directed by a boy to a mountain said to be covered with salt, and I saw a peculiar shaped cloud of a most beautiful silver colour, which assumed the proportions and appearance of a vast mountain covered with snow. Following its form downward, I became struck with the deep blue black colour of its base, and wondered if it portended another tornado; then as the sight descended to the gap between the eastern and western plateaus, I became for the first time conscious that what I gazed at was not the image or semblance of a vast mountain, but the solid substance of a real one, with its summit covered with snow . . . It now dawned upon me that this must be the Ruwenzori, which was said to be covered with a white metal or substance believed to be rock, as reported by Kavalli's two slaves . . .'

Thus it is that Stanley's name has become inseparable from the discovery of Rwenzori. In his autobiography he seals the matter firmly with his unequivocal statement that 'I have discovered the long lost snowy Mountains of the Moon, sources of the Albertine Nile . . .'

As the British brought stability to the newly created country of Uganda, a trickle of men started to explore this last great mountain discovery of the world. But surprisingly little impression was made on its exceptionally redoubtable fastness until the vintage year of 1906. It was then that the British Museum expedition researched the unusual high-altitude flora which will form such an important aspect of this book; and the British were followed by the Italian mountaineering expedition led by the Duke of the Abruzzi, a *tour de force* in which first ascents were made of all the major snow and ice peaks, solving the mystery of their complicated geography and giving us the mountain names that we know today. But the Italians covered only a few central square miles of a range that is some 60 miles long and 20 miles wide. The exploration of this thrilling wilderness – the secret Rwenzori which it is my purpose to reveal here – must primarily be credited to Dr Noel Humphries who, between 1926 and 1932, in a series of arduous expeditions, answered most of the remaining questions, leaving little for the first Belgian national expedition, which set out from the Congo in 1932, to fill in.

Since their existence was thus brought to modern Europe's notice, and up to the time of Uganda's independence in 1963, these mountains remained the province of a small élite of locally resident expatriates who for various reasons fell under their spell. Pleasing accounts of this golden period have been furnished by the published accounts of Rennie Bere (1955) and Douglas Busk (1957). Apart from these privileged few, most of mankind remained unaware of this range of snow- and ice-covered mountains in the very heart of Africa. Higher than the Alps, as extensive as the Bernese Oberland, uninhabited and pathless, the Rwenzori mountains are seldom seen by human eyes and only their more accessible parts are visited. The bulk of their extent remains the last part of the continent that is totally inviolate.

The reason for this lies in their position between the humid forests of the Zairean basin and the monsoon climate of eastern Africa, which causes them to be almost constantly obscured by cloud. As was the case with the early explorers, a man may spend months within their compass and never realise that they are there. Further, should he dare to try to penetrate their valleys, he soon discovers that the all-pervading cloud produces rainfall and

Below. Lobelia wollastonii, *in the early stage of flowering, above the upper Kitandara lake at 13,300 feet; the plant may reach 20 feet in height.*

drenching mists that are immutable. The consequences are a vegetation so strangely prolific that progress is almost intolerably difficult; a ground surface that is a soaking sponge; and a universal wetness that becomes icy cold as one ascends. Ironically, this exceptional natural phenomenon was more accessible and better known a generation ago than it is today, when sad Uganda, so beautiful but so ravished, is only just beginning to escape from her modern dark age.

The chances of war had taken me to East Africa in 1942 as a 22-year-old army officer, to take part in the raising of an African division for the war against Japan in Burma. After the war I spent twelve years in the Tanganyika Veterinary Department. In those days we were still victims of the delusion that the solution to the backward state of Africa lay in Western technology. It took a long time for us to realise our ignorance of the delicate balance of nature in Africa; a state of affairs in which almost everything one did was wrong, however beneficial it may have seemed at the time. It took even longer to appreciate that Africa was on a disaster course: that we were participating in nothing less than the irreversible disruption of what had been a stable, if fragile, ecology, developed over past millennia and discovered intact by the first European explorers; and that we were speeding its replacement by a chaotic instability.

The underlying cause was the increase in the human and livestock populations. Before these countries became independent, the Pax Britannica, with its medical and veterinary science, agricultural innovation and famine control, had taken the brakes off a potentially fertile population. Since independence there must be added to this problem the socially and politically motivated allocation of land for cultivation irrespective of the ecological consequences, and the concept, new to Africa, that land is a commodity to be manipulated for profit rather than used simply for subsistence. While in Europe and America it is the threat to wildlife species on the one hand, and on the other, the harrowing pictures of starving children that gain the headlines and rouse the emotions, the real problem is more profound. For we are talking of nothing less than the destruction for ever of the whole delicate covering of the earth's surface in these parts, leading inescapably to the demise of countless species of plants and animals, the drying of rivers, erosion of soil, desertification of the continent and the total impoverishment of the escalating human population.

If only out of self-interest, we in the Western world should be concerned about this calamity. The loss of any species impoverishes the environment of every one of us; the denial of a full life to any portion of the human population diminishes us all; powerful adverse climatic changes in one part of our small planet may disrupt our own more benign weather systems. And then there is the simple matter of beauty: why be afraid to speak of this? Africa was beautiful: year by year, under what are essentially Western-originated influences, she is becoming more ravished. This is a loss to the whole world and brings to mind W. H. Hudson's words in his childhood autobiography, *Far Away and Long Ago*: 'defence of the beautiful is God's best gift to the human soul.'

What was in the 1960's often dismissed as alarmist speculation has, unhappily, been dwarfed by the enormity of the disaster that we see all around us at the end of the 1980s. The prophets were right. I had met perhaps the greatest of these prophets in 1957, while I was enjoying a sabbatical break from Africa at Edinburgh University. At a small dinner party in an elegant Edinburgh town house I had been introduced to the late Professor F. Frazer Darling. Up to that time, as a tropical veterinarian, I had been an arch-proponent of the view that cattle were the means of turning the intractable thousands of square miles of East Africa's thorn bush and savannah into food for the benefit of the indigenous population. As the port circulated, I was astonished to hear the professor propounding a diametrically opposed view – that such a path would lead inevitably to the destruction of the continent. In the argument that followed I fear I may have embarrassed my hostess by my over-righteous refusal to haul down my flag in the face of this distinguished adversary. Was I not clearly right to be devoting my life to producing more milk and meat for those ill-nourished peoples? But as I walked home with the gentle-voiced professor and we shook hands and parted under the Edinburgh street lamps, I felt that I had somehow been touched. When I returned to East Africa, the opportunity arising, I switched from the headlong race of development schemes to more questioning field research.

During the years in which I soldiered and worked in East Africa, I made expeditions to many of the high mountain regions, not least Mount Kenya and Kilimanjaro. I thus had the good fortune to see these wonderful places in their pristine state before they were laid waste by the rising tide of human beings. In 1959 I at last kept faith with my Lake Albert vision of sixteen years before and made a long trip in the Rwenzori. I think it is true to say that during the whole of that expedition, the question of any environmental threat to the range never crossed my mind.

In 1961 I returned to work in Britain and twelve years were to pass before a research assignment took me back to Africa. I was shocked by what I found. I looked in vain for the beautiful forests and grasslands that I used to know. Now there was a wasteland of dust and erosion, while the forests that used to spread for many miles from the great highland massifs were sickeningly diminished. These

wonderfully diverse ecosystems, with their unique value as the only permanent water catchments of the region, were in the process of disappearing. Africa was committing ecological suicide.

Being at heart a mountaineer in the broadest sense of the word, I was particularly concerned at the threat to the beautiful high-altitude forests. Determined to make some record of them before they disappeared forever, and to try to discover the processes that were causing their destruction, I made a prolonged visit in 1979 to Kilimanjaro, Mount Kenya, the Nyandarua (Aberdare) range and the Mau plateau in central Kenya; but my plans to include the Rwenzori were aborted when the Tanzanian army invaded Amin's Uganda and that country became a no-go area. However, on my retire-

Opposite. *The view from our camp at the head of the Bujuku valley at 13,200 feet. Lake Bujuku lies at the foot of the formidable northern cliffs of Mount Baker. In the foreground giant groundsels are starting to put up their flowering spikes. The ground is covered with alchemilla.*

ment in 1984 I was able to make a prolonged journey through the whole length of the range, and later I was similarly able to visit the Virunga mountains further south. Although by definition it may be taken that at retirement one is older and less active, I did not see it that way. Rather I took the view that I now had unrestricted time in which to explore the mountains in detail, proceeding as slowly and loitering as long as I pleased. So it is that in the past few years, on four separate and long foot safaris I have been able to quarter the range. Even so, there remains plenty of virgin ground to tempt me back.

The outcome of my 1979 and 1984 journeys left me convinced that East Africa's remaining mountain forests are under imminent threat, with immeasurable consequences for the few permanent rivers of the region, not least the Nile. Rwenzori in particular lacks any form of statutory protection, and I decided that I should canvas the proposition that the Ugandan mountains should become a national park and a World Heritage Site. (The fifth of their area that lies in Zaire is protected by such status.) I decided that I should initiate this by setting my ideas down in illustrated book form. But photography has limitations for revealing the detailed beauty of flowers: I felt that the wonders of the afro-alpine flora called for a botanist who was an artist to do them justice.

I knew just a person. Christabel King is the daughter of the late Professor George King, one of the distinguished band of pioneer physicists who developed radar in the Second World War. The professor and his wife Priscilla were friends of my wife and me. Their daughter had graduated in botany at London University and then specialised in botanical illustration, becoming one of the visiting freelance artists working for the Royal Botanical Gardens. However, I could foresee problems. Christabel was not a mountaineer and had no experience of tropical travel. The region into which I was proposing to take her provides some of the most difficult conditions on the continent.

Special arrangements would be needed if she was to make a journey through the heart of the range and paint from life in each altitude zone. The first arrangement that seemed necessary was for us to take a third person who could manage our day-to-day arrangements while I gave Christabel the back-up she would need. I contacted David Pasteur, who had been an enthusiastic Rwenzori traveller, but unfortunately he was on the point of leaving for a long assignment in Sri Lanka. A little while later my phone rang. It was David's wife Ingrid, who was experienced in Uganda and had visited the mountains with her husband: couldn't she come in David's place? It was an ideal solution to my problem – a solution that was shortly compounded by their daughter, Elizabeth, asking whether she could come too, and bring her friend Caroline Massey? These two university girls had recently earned the Duke of Edinburgh's Gold Award and clearly were good expedition material. I shrugged my shoulders: at least I might find safety in numbers. Thus was born the 'British 1987 Ladies' Rwenzori Expedition', which will provide a thread for the account of the mountains that follows.

Opposite. *While the visitor's eye is understandably attracted to the strikingly photogenic giant groundsel species, numerous other medium and small species of this genus occur on the mountains, and they exemplify the variability which has permitted the development of the giant afro-alpine forms. This page from Christabel King's working folio shows four modest forms of* Senecio x pirottae *that she collected within a few yards of each other at about 13,600 feet as we descended from the Scott Elliot pass to the Kitandara lakes. They are thought to be hybrids between* S. matiroli *and* S. transmarinus.

Overleaf. *First light on the Portal Range (over 14,000 feet high). Most unusually, the Baker glacier, 11 miles away as the crow flies, can also be seen on the extreme left of the picture – an indication of our route up the Bujuku valley.*

292 Senecio x pirottae

Kitandava 8/8/87

all these collected
below the Scott Elliott
Pass at about 4120m
growing in grassy places
with Helichrysums & tree
Senecios

stems olive green
bladdish near tls.

white
hairs

dark reddish
tint on stems +
leaves only

2 cms

292b

No reddish
colour in lvs.
or stem

leaves all
about 108-109

292a

Flowers turning
orange, stem
some reddish
colour

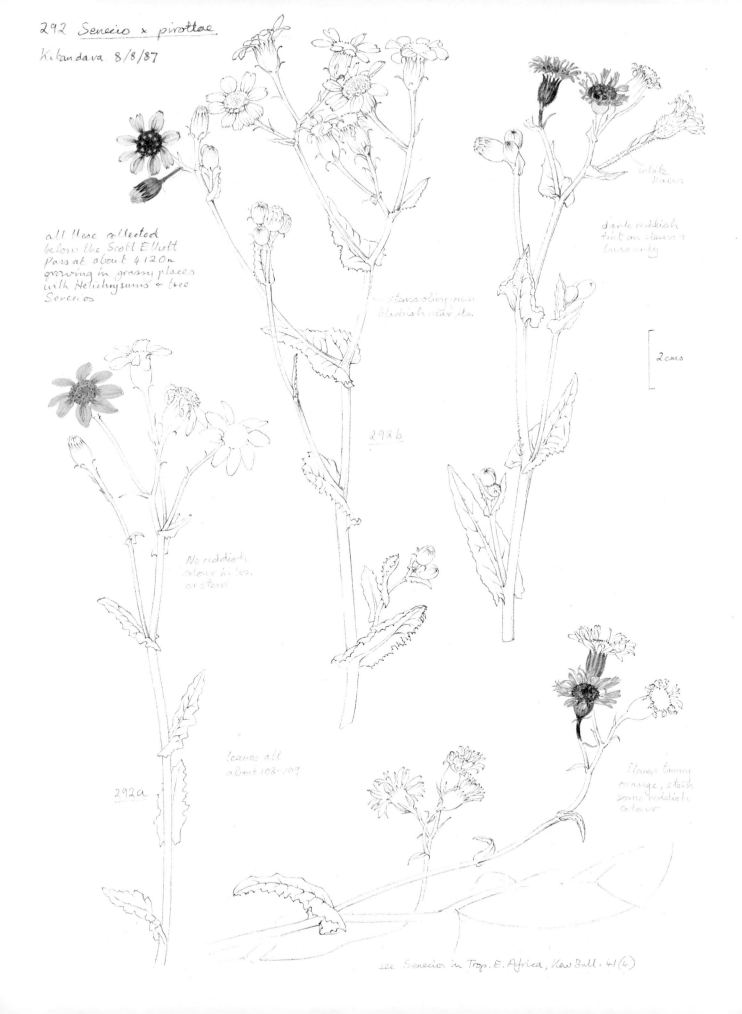

see Senecio in Trop. E. Africa, Kew Bull. 41(4)

TO THE INVISIBLE MOUNTAINS

The mountains – as yet invisible – were exploding before our eyes. The forbidding banks of dark cloud that filled the western horizon were breaking up into a frenetic artillery barrage of lightning strikes and thunder bolts. Africa was throwing her ferocious welcome at us. We knew what we were in for and within a few moments heavy isolated raindrops presaged a torrential tropical storm that flooded the soft black earth and bogged our truck down to the axles. Within a few minutes a body of local Bakonjo men, women and children materialised from nowhere and, with laughing good will, heaved us forward on to firmer ground, where at once the engine died. Perforce we sat immobilised, enduring the deafening drumming of the rain on the roof of the truck.

We had been established in our base camp at Nyakalengija, 5,000 feet up in the Mubuku valley, for three days, but had as yet seen no sign of the mountains. Reluctantly Ingrid and I had walked down the valley that morning and found a vehicle which took us to the small township of Kasese, 15 miles away, in order to make a courtesy call on the political head of the region and buy the quantities of dried lake fish that we needed for our porters. After an exhausting day of heat, dust and market argument, we set off on our return journey in the sultry late afternoon, and had just turned westwards on to the valley track when the storm had broken.

Our driver, a man of irritatingly sanguine temperament, dismantled the accessible parts of the engine – distributor, plugs, coil, petrol pump – mixing the bits indiscriminately in his hat, and repeatedly reassured us in the Swahili language, 'Don't worry. There's nothing really wrong – it's just that it won't work!' As the rain began to let up, repeated attempts at bump-starting by the now substantial crowd of well-wishers amply confirmed his words – it just wouldn't work. Dusk was upon us, and the smell of fish in the vehicle was nauseating; my companion and I shrugged our shoulders and resignedly set off to walk to our camp 8 miles up the valley.

As darkness fell and we trudged on through the softly rustling banana plantations, Africa, most temperamental of mistresses, cast her forgiving mantle about us. The now cool air was sweet with the after-rain scents of the earth; rejoicing frogs called out in piping notes from their gloriously unexpected puddles, and moth-like nightjars fluttered up from our path. Ahead of us, the black inscrutable mass of cloud that had been blocking the view up the valley began to undergo a transformation. From vapid shapelessness a noble structure of towering cumulus clouds emerged, and these became illuminated as if from within by a strangely awesome, bronze-coloured light, last vestige of the dying sun over Zaire in the west. Silhouetted against this, the mountains revealed themselves for the first time: still not the highest peaks, but the splendid Portal range which straddles the middle passage of the valley and forms the gateway to Rwenzori. As night finally took over, these rugged peaks became illumined by a flickering tracery of lightning, to a background of muted thunder. Rwenzori was giving us fair warning.

Over tea in our storm-tested mess tent, we related the day's adventures to our companions. Our reception by the political administrator had been reserved: after Obote, Amin, Obote again and now Museveni, suspicion was an ingrained feature of official life in Uganda. I had put all our party's passports on the table and these had been passed round the assembled staff, who took particular note of the photographs. I had read their minds: at my grey-haired age it was perfectly reasonable that I should have four wives – but the two youngest were only 19 years old! Wasn't that pushing it a bit? My explanation, that I had brought four ladies all the way from England simply so that one of them could paint flowers in the mountains, clearly lacked conviction. But Ingrid's smile won the day: the administrator gave us written authority to travel where we wished in the mountains, and we had parted with effusive handshakes.

I had not at this early stage yet grown much in the way of a beard, but I should mention here that travel in rural Africa is very much a question of striking good personal relationships with whomsoever may be in a position to help – or hinder – your progress, whether they be of high or lowly status. It is in this respect that to be of advancing years gives one an enormous advantage. One of the nicest aspects of African society is the genuine respect that is still paid to the aged. They are treated as truly senior citizens, and the most obvious way of announcing this status for a man is to display a grey beard. It goes without saying that the converse should also apply: visitors should always be sensitive to evidence of age in those with whom they are dealing, and treat such people with proper deference.

My companion for this night walk through Africa was Ingrid Pasteur, the only one of my party who had previous experience of Uganda. She had spent many years in the country as a teacher, and the alarms of our impromptu evening walk had caused her not the slightest

Below. *Slow train across Uganda. Expedition members from left to right: Caroline Massey, Christabel King, Elizabeth (Wiz) Pasteur and Ingrid Pasteur. The hurricane lamp and water bottles indicate previous experience of such travel.*

concern. The rest of our group, who now revived us with tea and supper, were her daughter Elizabeth (hereafter Wiz) and her friend Caroline, young university students to whom Africa was new, and our artist, Christabel King, who was the *raison d'être* of the whole expedition.

My previous experience of the mountains had endorsed the claims of the early explorers, that they present some of the continent's most intractable problems of travel. This is a consequence of their almost continual cloud cover; they are constantly subject to rain and snow, while at the same time the intense radiation due to their high-altitude equatorial position produces a vegetation of well-nigh impenetrable ebullience. My problem was that Christabel had never undertaken anything approaching my outlandish proposal before. How on earth was I to get her safely into the heart of the mountains and provide her with conditions in which she could produce fine work? As I have already described, the solution developed in the form of my support group of three other English women.

We all met in Kampala, the capital of Uganda, in early July 1987, equipped for six weeks in the field. The first thing we had to do was to cross the country from east to west by rail. The single-track line from Kampala to Kasese in the west was constructed in the 1950s for the purpose of carrying copper ore from the mine in south-eastern Rwenzori to the hydroelectric smelter at Jinja, the classic source of the Nile. The copper has run out and the mine is derelict now, but the railway still functions tenuously, a few unreliably scheduled passenger trains running each week, *sans* lights, *sans* water, *sans* lavatories, *sans* everything. It was into a single compartment of such a train that my party of five people and half a ton of expedition baggage were crammed for the overnight journey across the country. I had had doubts enough about committing my companions to the physical demands, discomforts and possi-

ble dangers of travel in the mountains, and these were not lessened by this initiation into African railway travel, with its intrusive machine-gun-toting soldiery and huge black cockroaches. I can only record that, whatever the inner feelings of my fellow travellers, they conducted themselves with admirable self-possession.

As the sun rose next morning our train began the winding descent of the escarpment of the western Albertine Rift valley, so called because it contains the inland sea of Lake Albert from which I had first seen Rwenzori in 1943. We eagerly scanned the western horizon for the great mountain range that we knew was there but, as had been the case with the earliest explorers, there was no sign of it.

In the chaos of the train's arrival at Kasese, we secured a utility truck and persuaded the driver to take us to our valley in the mountains. First we drove north along the tarmac route towards Fort Portal, through the inevitable road block of the National Resistance Army, and then west up the winding track to the village of Ibanda. Here our arrival was the excuse for an outbreak of euphoria of embarrassing proportions. In my case, veteran porters who had been with me on mountain journeys extending back nearly thirty years greeted me as a long-lost (and wealthy) relative at last restored to them. For Ingrid, the greetings were even more touching: she had previously visited the mountains with her husband David, and she and her daughter had to submit to the tearful embraces of all who remembered them, most especially from old John Matte, the *Mzee* (the word means 'respected elder' and is pronounced 'Mzay') patriarch of the Ibanda guild of mountain porters.'

In Ingrid's case the time lapse was some twenty years; in my case the immediate interval was only three years, since I had based my long 1984 expedition here. But I had first recruited porters in the valley in 1959 – twenty-eight years and a whole Konjo generation ago. One day in 1984 I had been walking up the

Above left. Mzee *John Matte, faithful past agent of the old Uganda Mountain Club, who more than anyone else has kept the tradition of the reliable supply of guides and porters going throughout Uganda's difficult years. He is now chairman of the recently constituted Rwenzori Mountaineering Services, a village co-operative dedicated to the proper management of guiding and porterage facilities.*

Right. *John's son Moses Matte, a leading headman of Ibanda, who has accompanied me on all my expeditions since 1984.*

valley through the banana plantations when I heard a cry from afar, and running down the hillside came a grizzled Konjo who threw his hoe to the ground and knelt to greet me. To recall old acquaintance is one of the joys of an African's life, and in greeting me in this manner this old veteran of my early expedition was not deferring to any imagined status I might have but simply making his statement about the marvellous felicity of this chance reunion.

'Remember,' he said, grasping both my hands, 'remember the great safari with *Mzee Kulekusengwa.*'

'Can you really remember me from so long ago?' I asked. 'We were both just youngsters then – why, it must be twenty-five years past.'

'How could I forget you?' he answered. 'You

are the only white man who eats rats.'

'Rats!'

I threw up my hands in simulated horror, but I knew what he meant. I had been collecting voles for the Coryndon Museum in Nairobi, and a bounty of 50 cents per head – or tail – had brought in more than I could handle.

Now our first task was to set up base camp, and to the disappointment of the villagers I insisted on this being at the end of the track, a further 3 miles up the valley at a pleasant spot called Nyakalengija, where we could organise ourselves and quietly acclimatise to equatorial mountain life.

Here we had straight away to solve the complexities of setting up our artist's studio. This large frame tent had been chosen with care: not only did it have to provide working space for Christabel; it also served as our communal mess and allowed on either side of a central work area private apartments for Christabel and Ingrid. More particularly, to allow Christabel to work under cover in the wet and gloomy conditions that so often prevail in the mountains, we needed a canopy, large translucent windows and a neutral grey canvas, so that her colour values would not be distorted. All these qualities we found in an excellent French tent that we now struggled to put up, to the perplexity of our Konjo friends; my colour coding of the rods and poles, which had been so elaborately worked out on my lawn in England, only added to their confusion. Certainly nothing like it had ever been seen in Rwenzori before, but still greater impact was made by the neat blue canvas ladies' loo, a kind provision from the Bromsgrove Girl Guides, which next had to be put up. Indeed I have little doubt that, in the African manner of remembering years by objects, our expedition will enter their history not so much for the women themselves, charmed by them though they were, as for the remarkable device dedicated to their modesty.

The campsite was pleasantly shaded by eucalyptus trees and was completed by our igloo tents, a log fire and a night watchman's den. There was nearby fresh water and delightful bathing in the rocky Mubuku stream, which here is spanned by an old wooden footbridge. The pretty green foothills, the ridges on either side and, when the clouds allowed, the splendid view of the Portal range across the valley above us, completed a setting that could not have been more pleasing. At this altitude, 5,300 feet, the climate is like English summer and it was my intention to spend four or five days here, to give our artist time to settle into her working conditions, and to organise loads and porters, buy rations and allow adjustment to safari life.

From this altitude upwards we would have no problems from mosquitoes or other noxious insects – except ants! These marching columns of fiercely biting soldier ants offer a minor hazard as far as the upper limits of the forest belt. Their pincer bites are painful although otherwise harmless. Our tents had not long been up before alarm calls were heard, the studio was invaded by marching millions and the unflappability of my party was tested to the limit. There is nothing to be done but evacuate the position and wait an hour or so, by which time not a single ant will remain. Every one of them will have followed their mysterious and compulsive urge to travel onwards: hence their common name, safari ants.

We had not been in camp long before little groups of children began to arrive with gifts: fowls, eggs, cabbages, tomatoes, onions and passion fruit. We soon established a proper

---

Opposite. *Gloriosa superba. Spectacularly beautiful varieties of this robust climber are found throughout eastern Africa at all altitudes. A Konjo child brought this specimen to Christabel King at our base camp in the lower Mubuku valley at about 5,300 feet, but I have found it growing to perfection at over 8000 feet on the Nyiragonga volcano in Zaire. The backward turned 'petals' vary greatly in colour; a striking variety is brilliant scarlet, hence the common name, 'flame lily.'*

Gloriosa superba

1 cm

payment system, to ensure continuing supplies and build up a reserve for our journey.

The preoccupations of organising our coming march into the mountains meant that we could spare little time for botanising, but Christabel soon got her needs over to the children and a trickle of touchingly beautiful posies began to arrive at her tent. These were put in water and the studio took on the air of a florist's shop. Field botanists may frown their disapproval of this haphazard system, but our priority was art rather than science, and in this way Christabel was able to make a good start to our study of the Rwenzori flora with these flowers of the lower valley, and work her way into the technique of painting in the field.

Meanwhile my concern was with logistics.

We were five Europeans with comparatively elaborate equipment and we were allowing for a thirty-day trip, on a basis of an average of three nights at each camp. We would need a mix of European- and African-style rations for this period, although a certain amount of fresh fruit and vegetables to boost these rations would be possible by relay parties. This all had to be packaged for porters, each man to carry an all-up load of 50 pounds, including his own personal gear. On top of this, we must of course take the rations for these men on the backs of yet more porters, for whom in turn yet more rations would be needed. It is a spiral, but it can be mitigated by sending down porters as loads are expended, to bring up further supplies. I had re-engaged my 1984 Headman,

Below. *The studio tent set up on a dry drumlin in the Mubuku valley at Kabamba, 11,400 feet. The ground cover is alchemilla and tussock grass, the background is helichrysum, groundsels and a copse of giant heath.*

Moses Matte, who is one of the many sons of Old John Matte. I also took on my previous guide, Peter Bwambale: these two make a first-class combination, and with a good deal of trial-and-error arithmetic in a school exercise book they produced a masterly plan of operations that proved entirely satisfactory. We would enlist thirty men to start with, and keep these in continuous service working up and down the valleys for the first two weeks or so; then we would progressively terminate their employment as the expedition turned for home.

To cover all this we would require rations for about a thousand man days. The hard core of the locals' rations is *muhogo* – cassava flour, locally grown and hand-ground by the women with large wooden pestles and mortars. This is the carbohydrate fuel on which the Konjo make their astonishing mountain journeys, and they need no less than 2 pounds per man per day. The protein to go with this is provided in the form of dried fish, *tilapia* and mudfish, caught by canoeists not far away in Lake George, then split open and either sun-dried or smoked. Each man expects one fish (the size of a large British kipper) each day, hence our trip to Kasese market for something like a thousand fish. To these basics must be added appropriate quantities of groundnuts, sugar, tea, dried milk, cooking oil, curry powder and plenty of cabbages, onions and tomatoes. Our shopping list was substantial and people were dispatched with wads of currency notes to the surrounding villages and markets, and the homesteads on the ridges. A trickle of shy Konjo women began to arrive at our camp. Each had on her back, as well as the inevitable baby lashed to her with swaddling, a large, beautifully plaited basket with out-turned rim, like an inverted bell, containing the *muhogo* – to be carefully weighed on our spring balance and paid for.

Other necessaries were a blanket and pullover for each man, against the bitter cold to come; half a dozen machetes or *pangas* – heavy-bladed bush knives for path- and firewood-cutting; five great aluminium cooking pans; sufficient

cigarettes and matches; and a supply of sacks for packing the loads. My bills soon started running into hundreds of thousands of Ugandan shillings, but such is the debasement of the currency that even at the unrealistic rate of exchange, my disbursements were measured in hundreds, rather than thousands, of pounds Sterling. I took a gloomy satisfaction in thinking of this welcome injection of cash into the valley economy.

Most of our business was transacted around an upturned tea chest in front of my tent in the shade of some flowering oleanders, where Wiz and Caroline kept us supplied with tea and coffee. Here Moses, Peter and elder statesman John Matte formed a more or less permanent committee. Earlier I had taken discreet opportunities to give each one of these veterans the gifts I had brought them from England. For the first two, this was a good-quality, heavy woollen long-sleeved pullover; for *Mzee* John, a pair of stout but smart leather shoes that would be used on the streets of Kasese, but certainly not allowed to be worn out on the valley paths! Our business mostly consisted of dealing with enquiries about supplies, and listening to beseeching applications to join the expedition. Old John's days of portering and guiding are over (and he fifteen years younger than me, I chafed him!) but he remains the unquestioned patriarch of the valley porters' guild, and is the chairman of the association they have recently formed, a co-operative called Rwenzori Mountaineering Services. John is a humble person, but of strong character, just and reliable, and by involving him at every stage in our planning, we effectively insured against subsequent minsunderstanding and default. Almost any potential dispute would be defused by my saying gravely, 'But did we not agree, in front of *Mzee* John himself . . .?'

Amongst our thirty porters we needed two specialists who would receive extra pay. The first of these had to be an English-speaking personal assistant for Christabel, a man of intelligence and sensibility who would stay with her constantly and carry her artistic and botanical gear. Erinesti Kitalibara was recommended for this important post and we could not have done better. Small of stature, with a wisely-humorous face – there was something of Puck about him – he had a gift for getting into scrapes that earned the good-natured laughter of his companions. He quickly grasped Christabel's needs and with his enquiring mind really took to heart her strange passion for the beauty of flowers. But many of these flowers would be far from easy to collect, so Christabel also needed a sleuth hound cum retriever who could penetrate the thickest vegetation and climb to the canopy for the flowers of the forest. For this we enlisted Sirasi, a tall, good-looking young man of fine physique, for whom no effort or risk was too much, provided only that he could bring us a perfect specimen.

Apart from these two specials, our men represented a typical cross-section of Bakonjo, ranging from young tyros, little more than boys, keenly anxious to enter the adult male world of portering, to grizzled elders. About half of them were veterans of my long 1984 traverse of the range from south to north, and amongst these I was pleased to see old Kajangwa, a time-etched, wise old pipe-smoking tracker and hunter – one of what is perhaps a fading generation of grand old men of the mountains. Another old hunter, Bathandika, as he squatted at our table, fumbled in the tattered and torn blanket that comprised his clothes and proudly produced for me a Nikon camera lens cap. He had found it when hunting on our northern route, and kept it for a couple of years for the pleasure of returning it to me personally!

Business between my committee and the local people is transacted in Rukonjo, the language of the Bakonjo people, little of which I can understand. My own communication with everyone except Moses and Erinesti, who speak English, is conducted in the Kiswahili language, which has been my second tongue for many years, a legacy of my East African army days and Tanganyika government service. Swahili and Konjo are both Bantu languages, yet they are distant, Swahili being a sophisticated blossoming of this great linguistic tree, while Konjo is a humble early shoot. The word 'Bantu' is often used as if it were synonymous with African negro, but this is incorrect, for there are many negro peoples who do not speak a Bantu language. Bantu refers to language: there are hundreds of such languages spoken throughout equatorial and southern Africa, and they all share similarities of structure and vocabulary. Because the Konjo speak what is considered to be an early Bantu dialect, they have rather unfairly got themselves stuck with the description 'primitive Bantu' by anthropologists. 'Primitive' in this sense should not be taken as pejorative, but as a synonym for 'early'.

A universal characteristic of Bantu languages is the custom of mediating word stems by the use of prefixes, as well as suffixes. Thus a single person is a *Mu*konjo, but plural, up to and including the whole tribe, is *Ba*konjo, while their language is *Ru*konjo and their country *Bu*konjo. In casual speech, however, most outsiders tend simply to use the root word *konjo* for most purposes, and that is what I shall do from now on.

These Konjo, whose company we were to share for several weeks and on whom we should be totally dependent in the remote fastness of the mountains, are of a dark chocolate complexion, with open frank faces, broad lips and nostrils and of only medium stature. They have a ready and genuine smile and a keen sense of humour. While they certainly are sturdy, they are usually spare and rarely carry any excess body weight. Their legs, which perform such wonders of balancing and weight-carrying over desperately difficult mountain terrain, can often be described as spindly. They live in homesteads of usually only one or two houses and a few small store huts, widely scattered and perched on the lower ridges of the foothills. The houses are rectangular, made of a double skin of plaited bamboo which is filled in with clay and roofed traditionally with grass or banana thatch. Almost the only concession to modernity that I have seen over the last thirty years has been the widespread replacement of thatch with corrugated iron. In the foothills, coffee is the main source of wealth; while on the plains, it is cotton.

The Konjo are a large and important ethnic group: there are probably a quarter of a million of them in Uganda, and substantially more than this in Zaire. They are an industrious and self-reliant people who, like the Swiss, have always been able to fall back upon their mountains. Thus they have largely stood aside from the turbulent comings and goings on the plains – whether of historical ethnic movements, or of the Arab slavers, or of modern politics and power games. To outsiders they may appear to have remained unsophisticated, but this is so only in the sense that they have not rushed to acquire the superficial trappings of Western culture – whether in the form of possessions or of ideas – that have in fact so often proved corrosive in Africa. They are, I think, immensely fortunate in this respect: the very knowledge that the mountains are there and that they are theirs – with that wonderful sense of eternity that the high hills give to mankind the world over – has provided them with an advantage over less fortunate peoples, and an anchor in the chaos that has at times engulfed Uganda and Zaire. Perhaps it is not stretching things too far to paraphrase Shakespeare and say that they enjoy a demi-paradise, a fortress built by Nature, which serves them against the envy of less happier lands.

They marry early – girls are betrothed by the age of 12 and marry at 13 or 14. Since they enjoy a healthy climate, sufficient food and pure water, survival rates are high by Africa's standards, and the population increase is all too buoyant. Most men probably aspire to polygamy, a second or even third wife being a proper ambition, but the majority either lack the resources for this or are constrained by membership of a Christian church, and prob-

ably at least two-thirds of marriages are monogamous. Dowry (bride price) is payable by the husband-to-be's family to that of the bride-to-be, and traditionally this took the form of fifteen or more goats. Only recently have any Konjo become cattle-owners, and then only to a small extent. Konjo women are of pleasing appearance, complacently relaxed albeit shy, and have a graceful modesty and nice manners. They never come as expedition porters, although they could do just as well as their men, so sturdily do they carry loads up and down the foothill paths. Massive stems of the great green cooking plantains, jars and jerry cans of beer or water, and large, finely plaited baskets of food crops and flour are transported only by the women – all this and baby too!

Apart from their position as home-makers and mothers, Konjo women are valued for beer-making and cultivation. The mens' part in cultivating is mainly that of pioneering and breaking new land; it is the women who manage the crop. Once the hills are left behind, bicycles begin to play a part in transport, usually ridden by men. A bicycle will carry three massive plantain stems, each weighing perhaps 40 pounds, as well as the rider. Until 1988, although there was a motorable track, there was no local motor transport based at Ibanda. Now at last there is a single small utility truck owned by Rwenzori Mountaineering Services. Other opportunities are very limited for men: a few artisans have skills in plank-cutting and carpentry, blacksmithing, basket-plaiting and so on. There are very restricted possibilities for small trading, shopkeeping and butchering; further afield on the plains there is cotton-growing and crop irrigation from mountain streams, limestone-quarrying for cement, and fishing in the lakes. A one-time major source of employment, the copper mine at Kilembe in the south-eastern foothills, is now defunct.

It is against this background of under-employment of young males that the porterage tradition should be viewed. The arrival of a major expedition such as ours is a matter of great satisfaction. A journey of this duration, well organised and well founded, quite apart from the welcome wage packet at the end, offers good food and clothing and cheerful company. There are opportunities for hunting, and side-lines in such things as making bamboo flutes, hoe handles and large wooden spoons, as well as the fun of getting away from family responsibilities. In short, for thirty lucky men, our expedition would be the high point of the year!

On the evening of 21 July we held our last council of war by the light of the pressure lamp in our mess tent. Our overriding priority was to obtain paintings of the mountain flowers and I had made it clear to my proposed companions before we left England that all other personal aspirations must be subordinated to this. The scientific botany of Rwenzori has been better studied than that of many other parts of Africa: ever since the British Museum expedition of 1906, the strange anomaly of this African alpine island on the Equator has attracted botanists. Lists of species have been published and, at least in the higher vegetational belts, these are probably comprehensive, since in terms of numbers of species the flora becomes less complex as you climb higher. But while these plants have been described, they have never been painted as works of art, and most of the world thus remains unconscious of these jewels in the crown of Africa. I felt that by bringing this strange beauty to the world's eye, we would also be powerfully emphasising the need to conserve it.

This priority determined our proposed route and itinerary. We needed to camp for as many days as Christabel required in each one of the changing climatic and environmental zones through which we would be ascending in our passage from the tropics to the ice cap. By following the only well-known path, which had been discovered by the Duke of the Abruzzi in 1906 and which has since become established as the regular route by successive parties, we would most readily achieve this. This is the route on which the Uganda Mountain Club in the middle of this century built five small huts – for the convenience of their members rather than for international tourism. These are now more or less derelict, and we were proposing to travel independently of them. But the huts had been sited as well as possible, in terrrain in which it is frequently almost impossible to put up even a small tent; these sites, then, dictated where we should camp to ensure our proper altitudinal sampling of the vegetation. Further, I knew that at each of these sites was a rock shelter, which would provide dry accommodation for our large party of Konjo.

As we squatted round the comfort of the Primus stove and Caroline made us a warm nightcap, while Wiz and Ingrid feverishly repacked our food bags for the umpteenth time, each of us must have been feeling that tomorrow would be something of a day of decision. Whatever one's well-intentioned planning, a journey must start with a single step, and with our very disparate ages and experience, each must have been wondering how we would match up to the demands that lay ahead. Most timorous of all must have been Christabel and myself: Christabel, because she had done nothing of this sort before and it would be a step into the dark; myself, because I had done so much of it, and I realised the absurdity of committing such folly yet again at my age. 'You are old, Father William,' I heard a voice saying, 'do you think at your age it is right?' Ingrid's self-control gave nothing away, and as for Caroline and Wiz, with all the buoyant spirits of youth, they must have been impatiently wondering why on earth we had been hanging around for so long.

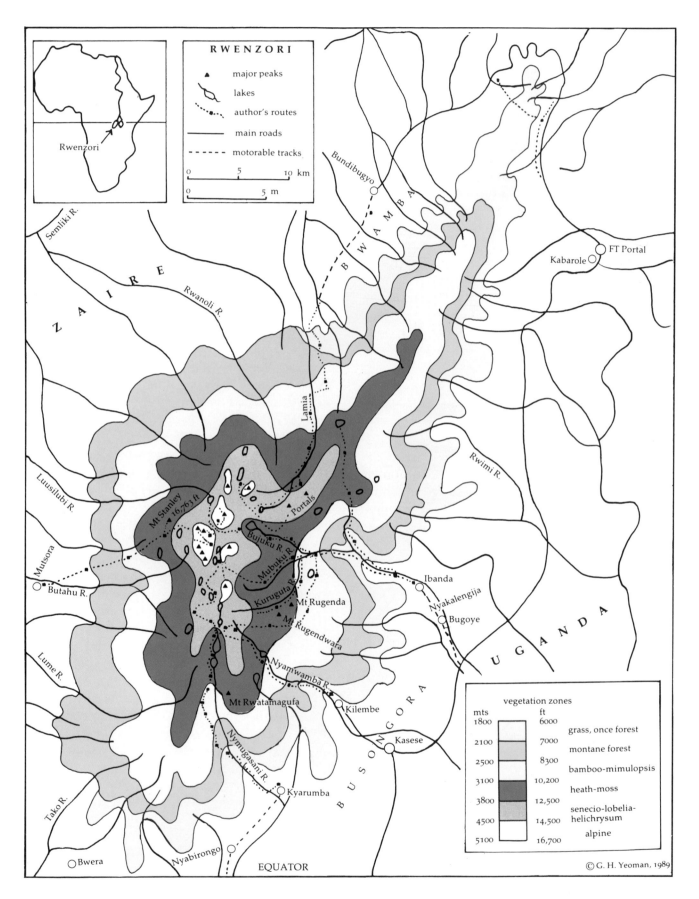

RWENZORI

- ▲ major peaks
- lakes
- ·■·■· author's routes
- —— main roads
- – – – motorable tracks

0   5   10 km

0   5   m

vegetation zones

mts	ft	
1800	6000	grass, once forest
2100	7000	montane forest
2500	8300	bamboo-mimulopsis
3100	10,200	heath-moss
3800	12,500	senecio-lobelia-helichrysum
4500	14,500	alpine
5100	16,700	

© G. H. Yeoman, 1989

Rwenzori

ZAIRE

Semliki R.
Rwanoli R.
Luusilubi R.
Mutsora
Butahu R.
Lume R.
Tako R.
Nymugasani R.
Bwera
Nyabirongo
EQUATOR

Lamia
Portals
Mt Stanley 16,763 ft
Bujuku R.
Mubuku R.
Kuruguta R.
Mt Rugenda
Mt Rugendwara
Nyamwamba R.
Mt Rwatamagufa
Kilembe
Kasese
Kyarumba

BWAMBA
Bundibugyo
Kabarole
FT Portal
Rwimi R.
Ibanda
Nyakalengija
Bugoye
UGANDA
BUSOZGORA

31

THE LAST OF THE NILE'S FORESTS

Striking camp next morning, packing loads and moving off, was a chaotic agony. The first day of a foot safari is always like this. There was a jostling, shouting crowd of a hundred or more people, to launch an expedition of thirty porters. Possibly my companions were expecting me to give an exhibition of authority, take charge and order everyone about. Nothing was further from my intentions. I knew that by any attempt of this sort I should simply make an ass of myself. I just kept up a low-key dialogue with my trusted headmen, such as was necessary to keep things moving more or less in the right direction.

Everything is carried by means of banana fibre. This is the flexible but strong material that can be stripped off the stem of the banana plant. Each load, whether a straightforward sack or an item of awkward equipment (Christabel's table and the ladies' loo were special problems), is adapted so that the main weight is taken by means of a headband made of the fibre, and the load is held firmly on the back by means of banana fibre shoulder straps. In ones and twos the loaded men disappeared and as soon as possible I sent Christabel off with Ingrid, while the girls shortly followed them. But there was no question of my starting until the last man had left and sundry stay-behind supplicants had been paid for minor services. The crowd evaporated and I was left only with old John, Peter and Moses. It was half-past eleven, far later than I would have chosen to set out on this first day's long march, and already I felt exhausted. Old John grasped my arm, bade me farewell and took his leave.

'Well,' I said, putting on my hat and grasping my staff, '*tayari twende* – let's get going.' But I was not in the least bit surprised when first Peter and then Moses, looking a little self-conscious, made excuses.

'You go ahead,' Peter said. 'I just want to pop down the road to say goodbye to my wife.'

'Yes,' Moses joined in, 'I promised to see my mother before we left. I'll follow straight on after you – I'll only be half an hour or so.'

I knew this was immutable custom: not just these two, but most of the others would by now be dispersed about the valley, consuming a last hot meal of *muhogo*, stoking up on roasted corn

Opposite. Traditional Konjo back-pack: an oval 'snow-shoe' lying flat against the back has head and shoulder straps of banana fibre attached to it. The load consists of one of my large bivouac tarpaulins and 40 pounds of muhogo *flour – all up, well over the standard air travel allowance!*

cobs and swilling last bowls of banana beer. Any idea of a cinema-style neat and organised departure, with a nose-to-tail line of porters striding off towards the hills, was naïve. In any case I was in a mood for my own company: this trip had been a year in the planning, and I felt content to take the first steps by myself.

'OK,' I said, 'But don't delay too much. I shall expect you both to catch me up by the time I reach the Mahoma stream.'

They departed and I was left by myself with my rather heavy thoughts, on the silent and deserted meadow that had been the hub of such activity for the last five days. I turned to face the hills and set out alone.

The little footpath I now followed through the last inhabited parts of the Mubuku valley was familiar. It winds amongst banana and coffee gardens and gives rise to narrow side paths that reveal homely glimpses of Konjo homesteads – clay, bamboo and thatch houses standing in bare earth compounds, with little store huts, racks for drying cassava roots and occasional hollowed-out logs for beer-brewing. Often the approach to the house is brightened by scarlet canna lilies or the spectacular foot-long white paper trumpets of moon flowers. Women at work spreading red coffee beans on mats to dry in the sun, or pounding cassava in mortars, called out 'Go safely' as I passed, and this was echoed in piping tones by the naked tiny tots who ran to join me for a few paces.

This is a heroic feeling, this welling of the spirit of the mountaineer as he notes the last evidences of felicity and home comfort and takes his final leave of valley-bound humanity. Soon the path narrowed, crossed the Ruboni stream by inconvenient stepping stones, and became enclosed by 12-foot-high elephant grass, making it hot and airless. I was grateful when the path steepened and climbed up the side of the Mihunga ridge, giving fresher air and fine backwards views over the country we were leaving behind.

Although most of the homesteads were now below me, this was still a zone of mixed cultivation and fallowed gardens (*shambas*). I was at the interface of the forest destruction which proceeds in Rwenzori not as dramatic clearance before axe and fire, but rather bit by bit, a process of unexceptionable nibbling – here for an irregular patch of yams, sweet potatoes or beans, there for a line of cassava plants or banana trees. It is an apparently unpurposeful and innocuous process that has none the less in my lifetime pushed the edge of the untouched forest a thousand feet up the steep hillsides and must inevitably, and with increasing speed, spell the loss of most of the broad-leaved forest. Here I was actually looking at the hard facts behind the proposals that I had been formulating with Dr Peter Howard of the World Wide Fund for Nature, for creating a national park for Rwenzori. Delicate though the higher afro-alpine regions of the range are, they are probably safe from human eco-vandalism. It is the lower broad-leaved rain forest that is the most rich in species and that is most at risk from Africa's ever-expanding search for new agricultural land. Its loss would be an aesthetic, scientific and economic tragedy and would rob the Nile of one of its last and most valuable resources.

The transition from the denuded foothills to the still-intact rain forest is irregular, and offers many delights, such as the massive, graciously leaved wild plantains, picturesque palm-like dracaenas, the fronded sprays of wild ginger, and wherever a stream runs, handsome tree ferns. I call this the aspidistra belt – house plants gone mad! Skirting the Mihunga ridge, the path drops close to the Mubuku torrent, where it is scarcely possible to resist stopping to sip the sweet chill waters. This is the last open space before entering the true forest, and it gives you an opportunity of seeing the vertical structure of the forest in cross-section, from humble ground-cover plants to swaying

Overleaf. *The rain forest of Rwenzori and the Mubuku stream, viewed during the ascent of the Nyabitaba ridge, at about 8,000 feet.*

canopy, plaited together by a tangle of vines and lianas. Climbing up from this point, I at last entered the cool damp of the forest itself, always for me a magical experience. For this is the climax vegetation of the continent, the ultimate biological complex, a living organism of untold thousands of intricately interdependent species of the vegetable and animal kingdoms. It is true that on these mountains the forest does not rise so tall as the lowland forests of the Zairean basin, but for beauty it is unsurpassed, and for the European visitor it has the advantage that the ambience is as delightful as an English summer's day.

Further, it is virtually free from noxious animals and insects. Almost, but not quite, for especially after rain one must keep an eye open for the marching columns of black or red shining soldier ants. If possible in this zone, I find it wise to make sure that a bare-footed porter is walking just in front of me. When, with an expletive of disgust, he and his load are suddenly elevated into the air, I know that I must watch out and step widely over the rippling dark column. I have already mentioned that there are no snakes here, but new arrivals can be forgiven for thinking there are, because one is entering the region of the giant earthworms. Not uncommonly these are found on the path; sometimes they are victims of the safari ants, who swarm around what must seem to them like stranded leviathans, slowly dragging them away to an unimaginable fate. Worms are difficult to measure, but a specimen that when contracted is 1 foot long, and when stretched, 18 inches, is not uncommon, and these may be as thick as your index finger. They vary in colour from the undistinguished hue of their European congeners, to a vivid iridescent purple or unwholesome green bice.

When I first made this passage in 1959 ants had not been the only problem, for the path had in places the appearance of a battleground, wherever herds of elephant and buffalo had their traditional and timeless walkways. Daily evidence of their passing was provided by the football-sized vegetable turds of the elephants and the dark greenish-black cowpats of the buffaloes. Further, the headman, old Kule, had beseeched me to take care in camp for fear of leopards. Now, less than thirty years later, these noble animals are almost extinct throughout the range, victims not so much of hunting in the mountains, as of the rising human tide that has denuded the once-forested surroundings and turned Rwenzori into a diminishing island.

These are not the only species that have been put under severe pressure in recent decades. Chimpanzees and Hoest's monkeys are now greatly reduced in numbers, as are the giant forest hog and the bushpig. The bushbuck, which is one of the most graceful of the forest antelopes, seems now to be almost totally absent from Ugandan Rwenzori, although I have seen them in the Virungas. Again, the basic reason for the loss of these species is not hunting but environmental destruction. They all require access to the lower montane forest for their survival, and it is this zone that has had to suffer the human population impact.

Some two hours after starting, as I began the descent to the Mahoma river which, rising in Lake Mahoma only a few miles away to the south-west, here flows into the Mubuku, I became aware that I was being followed. As good as their word, Moses and Peter, after finally severing the links, alcoholic and otherwise, that bound them to their homes, had silently caught up. I paused to let them join me.

'Jee! – did you leave all safe at home?'

'Everything was well,' they answered, and I could smell the *tonto* – banana beer – on their breath.

After crossing the Mahoma stream at about 7,000 feet, the path starts to rise more steeply up the densely forested Nyabitaba ridge. Amongst the finest of the forest trees are the symphonias, and here they grow particularly well. With handsome, silvery, straight, branchless trunks, they rise for 80 to 100 feet before spreading a glorious umbrella of dark green which is beautifully illuminated with masses of scarlet flowers. So characteristic of Rwenzori are they that I was determined to obtain specimens of their flowers for Christabel to paint, but neither Moses nor Peter would consider an ascent!

However, on a subsequent occasion I was able to return to the Mahoma with Peter Bwambale and Sirasi, our tree-climber. An hour of exhausting creeping about in the rain on the steeply uneven and tangled forest floor resulted in Sirasi at last blazing a fine trunk with his *panga*.

'This one,' he said briefly (for he is a man of few words) and at once began to roll up his sleeves and gird himself for the task. Cutting a length of vine, he tied the *panga* by its handle to his ankle with a 4- or 5-foot length, so that it would hang below him like a plumb. Then, spitting in each hand, he condescended to let Peter give him a shoulder up, and with a gruff 'Urghh' that spoke of formidable power about to be unleashed, he embraced the trunk with outstretched arms and legs and began the ascent. It was an impressive sight: the amazing musculature revealed beneath his dusky pastel skin pronounced man a true object of beauty. But I was petrified. I had no means of knowing what his margin of safety was, and I am afraid I was more worried on my own account than his. Suppose he fell and was killed, and I had to admit to a coroner's court that it was all because of my mad obsession for symphonia flowers!

Foot by foot he ascended by sheer friction, and at last reached the first branches that were swaying alarmingly in the canopy breeze. Scrambling into the latticed tangle, he hauled in his *panga*, identified a suitable branch and lopped it off to fall at our feet. Then with impressive sang-froid he lowered himself down the trunk, jumping the last 10 feet to receive my relieved congratulations.

The symphonia flower clusters were unusual in their beauty – scarlet globes like cherries, well named *globulifera*. Knowing that they would not be amenable to pressing, I took a series of close-up photographs and then immersed them in spirit in a plastic collecting jar I

had brought for the purpose. Sirasi watched my careful processing with approval and then called my attention to something I had not seen: on the very branch that I was photographing was a handsomely camouflaged reptile – a large three-horned chamaeleon dressed in green and gold, a kind unique to Rwenzori. There are several species of chamaeleon special to the mountains – I have found three or four kinds. To me they are always a happy discovery, but Africans do not like them. They play an unfavoured part in Bantu mythology, and the people superstitiously prefer not to handle them. I had no such reservation and repeatedly posed my scaly friend until I was happy that I had his portrait.

But on the present occasion of our first day's march up the valley I had to admit defeat over the symphonia and be content with the other flowers of the forest. These included large pale pink balsams, purple hibiscus, white button-hole begonias and spectacular lilac-coloured thunbergiopsis. Some way up the Nyabitaba ridge a landslide had sheared an opening in the forest and left a pleasant vantage point with fine views up and down the valley, and here I found my companions taking lunch. I had made a secret vow never to leave Christabel out of my sight on the march, and from now on we travelled together, leaving Ingrid and the two

---

Below. *The three-horned chamaeleon of Rwenzori, on symphonia leaves at 7,500 feet.*

Symphonia globulifera

girls free to go ahead and set up camp; this remained our *modus vivendi* for the rest of the expedition.

The higher part of the ridge opens out into bracken, and charred tree stumps tell that the bracken has colonised old forest-fire scars. Such bracken-filled scars are common all round the mountains: for all its persistent rain, there are years when the undergrowth can become dried out and then, so dense is the mat of flammable material, fires can spread with great ferocity. The bracken is the same species that we have in Britain, and it is common throughout all the highlands of East Africa. Beyond the bracken we re-entered the forest on the now narrowing ridge, and a steep few hundred feet brought us on to a veritable knife edge from which, looking back on a clear day, one has a fine view of the Rift valley plain of Uganda, where Lake George lies like a silver mirror – the country through which Stanley had passed so painfully more than once, a hundred years ago.

Here we found ourselves in the podocarpus forest that is characteristic of the 8,000–9,000-foot contour. These podo trees are handsome evergreen conifers of the yew family, but rather than needles they have thin linear leaves which, with their tiny, mauve, bloom-dusted cones, give the trees great artistry: when bathed in sunlight (a not-too-common occurrence) they give off a smell evocative of English yews.

Hot and dry-throated as we were, I was able to cheer Christabel by predicting that the angle of the ridge would decrease and that camp was not far off. This last part of the ridge is in fact a delightful airy walk in fine weather such as we

Opposite. Symphonia globulifera. *The symphonia trees tend to occur in clusters and are characteristic of the montane forests of the equatorial Nile catchment at around 8000 feet. They stand out in the canopy with their masses of scarlet blossoms. Christabel King gives us an impression of the 100-foot specimen that Sirasi climbed, on a steep hillside above the Mahoma torrent, and reveals the strange cherry-like flowers.*

were now enjoying. Invigorated by sweeping views over the top of the forest canopy filling the valley 500 feet below, we came at about sunset to the first rock shelter of Nyabitaba. The camp is a little beyond, under fine podocarpus trees, and here we found our tents up and the kettle on the boil. Our first day's march was successfully accomplished.

Nyabitaba is one of the best campsites in Rwenzori, for it is dry underfoot, delightfully dapple-shaded by the podo trees, and it provides impressive views, across the deep Mubuku gorge, of the southern Portal peaks. This range, which, when it was clear of cloud, we had admired sideways on from our base camp, we now saw close up, end on, towering 6,000 feet above us, a monumental battleground of clouds which at night flickered and reverberated with lightning and thunder. The name 'Portal' is interestingly ambivalent. Sir Harry Johnston, with characteristic poetic flare, suggested the name because this range within a range did indeed 'serve as awful portals of the snows', but he combined this with a dedication to Captain Raymond Portal, whose ineffectual journey across southern Uganda in 1883 – part of the British Mission under his brother, Sir Gerald Portal – did not even reach the spot where Fort Portal now stands.

Nyabitaba is dry because it is on a ridge: only on the old Belgian route from Zaire in the west do you find similar dry ridge sites. The ridge at Nyabitaba is a vast glacial moraine, one of the largest moraines in the range, and it reminds us forcefully that the present glaciation of Rwenzori, splendid though it is, is but a token relic of the glory that existed in the last African ice age of 15–20,000 years ago. Then the whole central valley network – Bujuku, Murugusu and Bukurunga – were rivers of glacial ice which, combining and forcing their way eastwards, were joined above Nyabitaba by similar ice flows from the combined Mubuku and Kuruguta. It was this massive ice jam that deposited the material on which we were now camped, and the ice tongue had continued far down the valley up which we had passed, to terminate well below the 7,000-foot contour, not so far above our base campsite. Everywhere in our journey to come we were to see evidence of this ice sculpturing: dramatic U-shaped valleys, entrancingly beautiful cwm tarns and lakes, invaluable ice-undercut rock shelters, lugubrious bogs and exposed terraces and precipices. Dr Henry Osmaston, doyen of modern scientific Rwenzori travel, co-author, with Ingrid's husband David, of the only pocket guide to the range, has opened our eyes to these often awesome features and greatly enhanced our pleasure in understanding the mountains. But for the mountaineer, this pleasure is sadly tinged with regret for things past. Grand though Rwenzori is, how infinitely more splendid, and what a phenomenal wonderland it must have been 20,000 years ago. Now we know that, if the rate of ice depletion we have observed in our lifetime continues, or indeed increases under the greenhouse effect, the snow-capped sources of the Nile will only remain snow-capped for a generation or two to come.

We spent four days at Nyabitaba and enjoyed it immensely. Every night was rocked by ear-shattering thunder and brilliant, often almost continuous lightning – the same artillery battle that we had seen from the plains; but every dawn was dazzling in its clarity, bringing glorious summer days. Christabel was hard-pressed to keep pace with the specimens we brought her, but she is a quick worker and her folio began to bulge. Her aim was to make working drawings from life, with sufficient watercolouring and notes to enable her, with the help of preserved specimens and photographs, to produce her finished plates when she returned to Britain. Her hero in the world of botanical illustration is Ferdinand Bauer, an Austrian botanist and illustrator, who in the eighteenth century had co-operated on the Oxford *Flora Greca*, travelling and collecting in Greece and the Levant. He used a system of coded colour charts, such that he could just

annotate his sketches with numbers, quickly recording the correct colour values. Christabel had devised her own version of this, and we were deeply impressed by her workmanlike approach in this unusual situation. We had with us a copy of Agnew's *Upland Kenya Wild*

Flowers, the best guide available to us, and Ingrid worked assiduously, producing provisional identifications and writing up the official Kew Herbarium labels before committing the specimens to the presses.

I have mentioned that the ridge provides a fine aerial view over the forest canopy, and it is one of the best places to see Rwenzori turacos,

Above. *The upper rain forest in the Mubuku valley, viewed from the Nyabitaba ridge at about 8,000 feet.*

41

which are such a thrilling feature of the forest. Medium-sized, about the size of an English jay, they are at first likely to be heard rather than seen – an evocative clicking-cawing as, inconspicuously in their flocks, they search the podo trees for the little cones they so dearly love. Even if you should pick them up through binoculars, silhouetted against the sky, it is hard to appreciate the glorious plumage – a *mélange* of iridescent green, purple, blue, scarlet and violet. But if you look out from the ridge, with luck you may see a group of them take flight at eye level when, if the sun is at the right angle, flashes of flame will be transmitted through their scarlet flight feathers. Also in this area you may see a school of black and white colobus, a sub-species of these most handsome primates which is peculiar to Rwenzori. They have jet-black bodies with flowing white capes and tails which they use as wings to plane through the canopy.

Our party of thirty porters, who had mostly scattered to their homes as soon as they had shouldered their loads at Nyakalengija, had as predicted all homed in on Nyabitaba, those who had not set off before me slipping past in ones and twos during the afternoon, quietly coughing to beg me to allow them to squeeze by on the narrow path. Now they had taken up quarters not in the first rock shelter but in a larger and better one a couple of hundred yards beyond, a little below the crest of the ridge. This shelter is formed by a giant erratic boulder, and could provide room for a larger party than ours. By the time I got round to visiting them it was nearly dark, and they were settled round two blazing fires of podo timber, squatting on logs and watching the labours of the cooks, who were stirring and dolloping the 60 pounds of *muhogo* that was to be their supper. This is a traditional place for making the big spatulate spoons which they use for this purpose, and also for making shafts for hoes that they will take down later to their womenfolk. It is also here that the indispensable staves that we were each to carry for the rest of the trip

would be cut – light, coppiced stems some 5 feet long, of a hickory-like wood that is flexible but almost unbreakable.

It was a pleasing scene, the men's animated faces illuminated by the warm firelight, laughter and talk constantly passing to and fro. As I approached, holding my hurricane lamp, I called out softly, '*Huti, huti*' – the Konjo polite, 'May I come in?' – and a chorus answered, '*Pole, Mzee, pole-karibu keti*', commiserating with me on my journey and begging me to come and sit down. As I squatted amongst them I noticed two dogs.

'What's this?' I said to Moses, with mock severity. 'I don't remember enlisting dogs – did you write down their names? And what rate of pay was agreed?'

My corny wit was greeted with laughter and knee-slapping.

'Jee! As if dogs have names – ha! But they will get their pay all right. Don't we give them the *muhogo* pots to clean out, and the burnt bits? La! How do you think we can hunt without dogs? But they will eat the hyrax guts and bones.' (Hyraxes, rodent-like 'rock rabbits', are the most common quarry of Konjo hunters.)

The dogs were nondescript mongrels, with enough rib showing already, and they certainly wouldn't put on any weight on this trip – but they lay there, shining black eyes reflecting the firelight, with their noses between their paws, supremely content to be getting a little share of the fire which, flaring up, erratically illuminated the rugged roof of the cave.

On our second afternoon at this camp some strange porters started coming up from below and I was told that we were to expect a singular

Opposite. *My Konjo companions have a marvellous flare for turning unpropitious places into snug camps. The slanting overhang is typical of many such rock shelters. The cook of the day is stirring the* muhogo *dough with a large home-made wooden spoon. The man turning towards the camera on the right-hand side is Kajangwa, our path-finder and hunter.*

honour – Father Clichet was on his way. In answer to my enquiry, Moses explained to me that this ancient priest was one of the French White Fathers at the famous Catholic mission at Mbarara, a town about 80 miles away to the south-east. He was said to come to the mountains every year, and it was reputed that he had done this no less than twenty-seven times! Within an hour he was with us, a man so weather-worn that I found it difficult to guess how old he was, but such was his ascetic Mahatma Gandhi-like appearance and his bent gait over his staff that he could have passed for a lively octogenarian. Ingrid made him a cup of tea while we conversed as best we could in a mixture of English, French and Swahili.

Yes, all they said of him was true. Three hundred and fifty-five days a year he laboured to serve God in the dust and heat of the plains: was it not good, for the other ten, to visit Him here in His provenance? The priest raised his eyes with humble sincerity to where the clouds hung above the head of the Bujuku valley. This was his solace: perhaps he would never see France again (the White Fathers have a demanding discipline) but the mountains had always been his consolation.

I was at one with him, however disparate might be our views on religious matters. Mountains to me are infinite resources for the re-creation and refreshment of the human spirit.

'I see you are like me,' he said, laying his hand on mine, 'You too come all the way from England, to recharge your batteries.'

I told him of my fears for the forest.

'Of course you are so right,' he said, 'We are helpless in face of the deluge: soon there will not be a tree left in Uganda. But isn't that all the more reason to trust in God's providence?'

'God will only provide', I answered, trying to sound neither sententious nor contentious, 'if we ourselves take up the task. All I can do is to press the case for a national park so that others, in generations to come, can also come here to recharge their batteries.'

Our philosophical discussion was interrupted by the arrival of three charming but weary young Frenchmen, short-service lay volunteers at the mission. I could see by their exhausted looks that the worthy Father did not have my ideas about moving slowly uphill. Wiz, Caroline and Christabel hastened to make more tea, and the young men's astonishment at having stumbled into this English tea-time world of charm and beauty in the mountains was delightful to see. Later that evening one of them came shyly to me and confessed that he was gravely disabled by tropical diarrhoea: I gathered that the Spartan Father held little truck with modern medicine.

'Your headman tells me you are a doctor,' he said. 'Please can you help me, or I fear I may fall down on the way tomorrow?'

'Certainly I am a doctor,' I said, reaching for my medical pack, 'but mostly of cows – a *vétérinaire*. But do not fear: when it comes to diarrhoea we vets can beat human doctors any day. I'll promise you a good night's rest and strength for tomorrow's march.'

My remedies did not fail: next morning he looked a new man, and when some days later we met the party descending, he embraced me warmly as one who had saved his life!

THE VALLEY AND THE SHADOW

As we descended from our high ridge camp at Nyabitaba into the deep forest-filled trench of the Mubuku valley some 500 feet below, the roar of the river in spate filled me with apprehension. Here we were at the confluence of the Mubuku with its major tributary, the Bujuku, and after the nights of rain, it was quite likely that the river would prove impossible to cross. The day before, Caroline and Wiz had gone ahead with a reconnaissance party, cut a tall tree and wedged it across the torrent in an attempt to make a handrail, but it fell short, and did not help over the powerful central race. On seeing the surging waters, the porters dumped their loads with expletives of disgust. They debated running our climbing line from bank to bank, but in the end settled for unprotected wading through the chill waist-deep cascades. Bowed under their bulky loads, and with bare feet the men sturdily picked their passage through the torrent with exemplary sang-froid and cheerful bursts of laughter, straining their bodies against the flow with their long poles.

Christabel took only casual notice of all this activity and appeared to banish the thought of the coming crossing by giving her attention to the vegetation. The shady gallery forest beside the river, with its creeper-hung trees and cool, moss-covered rocks, is one of those special places that make you cry out in your heart, 'This is the Africa I dreamt of as a child!' – even though you know that most of Africa is hot and dusty, the antithesis of this scene of lush vegetation, rushing pure water and cool air.

For the moment Christabel delighted in the elegant trumpets of Emin's bell flowers, hanging from an ancient tree trunk like Victorian Christmas lanterns, while from the damp earth below were growing the rather sinister flowers of the snake's head arum, with its cobra-hooded green- and cream-coloured sheath, reaching 3 feet tall. These, and the nearby clumps of amaryllis with pendent clusters of flame-red flowers, were high on my list of requirements and we set to with sketch pads and cameras, determined to do them justice.

Eventually we could no longer postpone committing ourselves to the waters. Our three female companions had already crossed with evident aplomb, but I was not looking forward to the experience. Three years ago at this spot, with the river in similar spate, I had been swept off my feet and hurtled down the water chute. I usually carried my cameras round my waist on a Camera Care belt, but I had slung them round my neck for the crossing. As I surfaced from the ducking, I had the presence of mind to seize the belt and sling it with all my strength to old Baloko, one of my porters, who had been anxiously preceding me like a sturdy St Christopher, bowed on his pole. He caught it with his free hand like an expert wicket-keeper. In the same split-second, my guide, Peter Bwambale, had thrown himself into the torrent, seized my arm and dragged me floundering and shamefaced to the safety of the rocks, where I was greeted by the delighted clapping and dancing of my Konjo audience. It was just the thing to make their day!

With such gloomy prognostications in mind, I now insisted on our photographic and art equipment being sent across first. Then, stripping off as far as decency permitted, we in turn surrendered ourselves to the care of the porters who rather ignominiously bundled us from man to man through the flood. In this way we all found ourselves wet but safe on the far side of the river, the last to cross being our desperately anxious dogs which were carried on the necks of the men as lambs are by shepherds.

The river crossing was just a prelude to the day's exhausting work. We were only at the beginning of the Bujuku valley and our path now took us up the steep valley side through a well-nigh impenetrable tangle of mimulopsis. This plant is the archetype of a range of densely

growing, straggling, semi-woody vegetation that extends over immense tracts of the mountains. Its lower stems form a mat-like covering on the ground while its leafy vertical spikes are often tall enough to arch over the walker's head. This wall of vegetation is usually saturated with water from mist and rain, which makes one's passage even more disagreable as one is showered at every step. The porters slash a path with their *pangas*, creating an unpleasantly slippery mattress of stems over the waterlogged ground. As we progressed slowly through this green world, we could be forgiven for not admiring the mimulopsis flowers which, while often not profuse, are indeed beautiful when freshly opened – a crepe-like trumpet of cream, splashed with orange.

The attraction of Rwenzori flowers is often that of surprise rather than profusion. In the valley tangle it is easy to miss the astonishing fire-ball lily: on a single stem, about 1 foot above ground level, and before any leaves have appeared, are carried a hundred or more delicate scarlet stars with tiny yellow anthers, forming an exploding globe of fire 6 inches across, reminiscent of a fibre-optics display.

Other less attractive plants lurk amongst the mimulopsis. Stinging nettles, whose leaves are insufficiently distinctive to make themselves obvious to the uninitiated, are a feature of this altitude belt. Their sting seems agonisingly worse than that of our common European nettles, but unlike the latter the pain is short-lived. You must simply ignore it and concentrate on the other miseries afforded by the path. As you climb higher, banboo starts to mix with the mimulopsis and its criss-crossed, half-fallen stems and matted spillikins underfoot are a tiresome obstacle.

Opposite. *Crossing the Mubuku torrent is always problematical. Here two of the men are trying to fix a tree as a hand rail. Plans are in hand with USAID funds to re-instate the 'Busk' suspension foot bridge which was swept away many years ago. The stream here abounds in brown trout.*

At these altitudes – the day's march would take us from 8,300 to 10,900 feet – even though the sun is obscured by low clouds and drenching Scotch mist, the penetrating solar radiation of the Equator combined with one's energy output can make the going pretty warm and put the traveller into an uncomfortable sweat. Yet, because of the denseness and pervading wetness of the vegetation, it really is best to wear an anorak with waterproof trousers and gaiters. From previous experience I had advised my fellow travellers to fit themselves out entirely with modern Gortex-type 'breathing' waterproofs and our consensus was that we could not have done better.

As we forced our way slowly and painfully through this vegetational barrier, I was full of anxiety about Christabel. I knew little of her physical capability and less of her psychological strength. She is a highly educated woman, who had done virtually no camping or hill-walking in her life. What keeps the climber going, like the long-distance runner, is the sheer dogged mania of wanting to achieve. Did she possess this, or had my whole plan been irresponsible? Were we heading for fiasco? I found myself surveying her slight figure, sometimes walking behind her so that she could set her own pace, sometimes in front, so that I could set it even slower. Whatever she was thinking, she gave nothing away and never a word of complaint passed her lips.

Equally I had misgivings about my own capability. When you are young, even if you are not fully fit at the outset of a trip, your body adjusts; but in your sixties you are not so resilient, and I had to face the fact that it might be my own fallibility, rather than that of my companion, that could bring the expedition to a premature conclusion. After several hours of purgatory, these melancholy thoughts were reinforced by our arrival at the sombre cliff-face of Kyemera: our path was to pass along the foot of its rocks. These cliffs are the first tangible evidence of the stark glacial origins of the valley. I must have passed them a dozen times, always with trepidation in my heart, for it was at this spot, on my first visit to the mountains, that I had nearly died.

In 1959, after driving for three days with my wife and children from our home on the shores of Lake Victoria, I had reached the mountains a week before my climbing companion, John Newbolt. Leaving my family at base camp, I made a preliminary solo trip up the Bujuku with a party of Konjo, in order to place a depot of supplies at the valley head. My headman was Kulekusengwa, a rather morose but reliable Konjo who, although some ten years older than me (and that made him about 50 years old), was still capable of the extraordinary speed of ascent typical of his tribe. Foolishly, instead of insisting on him adopting my usual more measured pace, I forced myself to adopt his: I suppose I wanted to make my mark among my new companions. This worked reasonably well until, on the second day, when we were approaching the Kyemera cliffs, I found myself going badly and had to call for a rest. Kule and I sat down side by side on some stones with our backs against the cold cliff.

As I sat there I was slowly overcome by a creeping paralysis. At first I realised that I could not move my legs, and then my arms contracted tightly against my chest, fingers curling inward like claws. My upper chest, my neck and then my face became numb, so that I was incapable of speech. I desperately wanted to appeal to Kule to help me, but could not. He just sat staring at the ground, his head buried in his hands, unaware of my plight. By now, only the lower chest muscles, with which we breathe, were still functioning. But my mind remained clear: with a mounting sense of terror I realised that they too must shortly fail, and death would be inescapable. My thoughts crystallised on my wife, less than 10 miles away, and on the appalling consequences to my family – I could not even leave her a message of love.

At last Kule turned to look at me and I saw the alarm in his eyes. He grasped my claw-like arms and urgently tried to massage and straighten them. To this day I can recall the softness of his cold fingers. As my voice began to return, I was overcome with thirst; but when at last I could get out the one word *maji* – water – I still could not grasp the cup and he had to hold it to my lips. Inch by inch movement slowly returned to my body, while the black shadow over my mind began to lift and I was overcome by a euphoric sense of release. Half an hour later, I was on my feet and stumbling forward.

Nearly thirty years had elapsed since that day and on our present trip I sought out Old Kule. Now over eighty years old, he lives the life of a patriarch in a remote valley amongst his coffee gardens, surrounded by his daughters and grandchildren. His melancholy face seemed almost unchanged; his eyes filled with tears as he recognised me, and those same gentle fingers grasped my forearms in the greeting of special affection. We sat in the shade of his banana fronds and talked of times past, while his daughters hospitably slaughtered a goat and busied themselves with the preparation of baskets of *matoke* – steamed bananas – and *muhogo*. When I mentioned the incident at Kyemera, an anxious look crossed his eyes.

'*Bwana*, you know there are spirits at that place. I thought that day they had come to take you from us.'

In retrospect, I believe that this close call was a cramp due to hyperventilation. In my Konjo-style ascent I had been forcing my respiration and had overdone it. The incident impressed on me indelibly the virtue of slow progress in mountain-climbing. The best-known of Swahili aphorisms is *Haraka, haraka, haina baraka* – haste, haste, it has no blessing. Alex Taugwal-

---

Opposite. Mimulopsis elliotii. *People who have climbed above the forest belt on Rwenzori will have mimulopsis engraved in their minds. Vast tracts of this tangled shrubby weed form a difficult barrier. None the less the sparse flowers are attractive when freshly opened, although they soon fade to a drab grey. There seem to be several species.*

Mimulopsis elliotii

der, an old Zermatt guide, expressed the same wisdom to my brash young self years ago, as we left our hut before sunrise to make a glorious traverse of the Rimpfischorn: 'Monsieur Yeoman, you must please learn that the *slowest* way up a mountain is the quickest!'

To people who are new to these matters, I would say beware of a companion who strides out urgently, and more especially, of a leader who strides ahead of his party – it is almost certainly a sign of inexperience.

These reflections were going through my mind as Christabel and I sat side by side at the foot of the Kyemera cliff, sipping our lemon drink and nibbling groundnuts. I did not tell her my story but I lost my fears about her. Clearly she was going to make the day's march without difficulty.

We soon moved on, leaving behind the shadows of the past, and to our relief, the detestable mimulopsis began to give way to bamboo and the first hypericums – straggly trees with beautiful, golden, cup-shaped flowers, the St John's wort or rose of Sharon. Scattered hagenias made their appearance, handsome, craggy and picturesque trees, like giant unkempt versions of bonsais, and in one of them we saw a school of the blue monkeys that are a pleasing feature of Rwenzori and a favoured quarry for Konjo hunters. As we advanced, our path thankfully levelled off; the valley narrowed darkly, rain began to fall and in the deepening gloom we came to the derelict hut of Nyamuleju at 10,900 feet. This stage from Nyabitaba generally takes four or five hours: we had been travelling for seven and a half! But how wise we had been, for we came in strongly and I was left with a sense of confidence in our ability to tackle the demands that lay ahead.

There is a fine rock shelter at Nyamuleju, although it was a tight squeeze for our thirty-two men: an almost horizontal roof keeps it dry and comforting smoke was emerging from two fires of massive heath timbers. Good though the shelter is, the Konjo do not like it: it is said to have an association with death, and with my

memories, I could hardly scoff. For my companions conditions were less comfortable. The old hut was derelict beyond use, and so steep is the valley side – awesome cliffs above and a desperate tangled drop to the river far below – that it was only with ingenuity that we could squeeze the two igloo tents on to uneven platforms, after a brisk slaughter of the rampant clumps of docks with which they were overgrown. There was no question of putting up the big studio tent, and I joined the porters in the cavern, where they made a bower for me out of a tarpaulin.

Setting up camp on a bad site in rain is a test of moral fibre, but the problem was overcome

Below. *Kulekusengwa, who was my guide in 1959; photographed with his great-grandson in 1987.*

with universal good manners. This was the first of many occasions to come when Ingrid proved her unflappable ability to organise quick hot meals despite difficult conditions. As the sun set prematurely behind the darkly crowding mountains, I recalled that on my first journey I had climbed up the moss-covered rocks above the camp and been rewarded by spectacular views of the snow peaks and glaciers – Mounts Baker, Stanley and Speke. The rain letting up, the two girls and I attempted to repeat the experience. It was a mistake; dangerous flounderings over slippery rocks on water-saturated moss revealed nothing more than the steep valley sides disappearing into uncompromising dark clouds. Bujuku would concede us nothing.

Most of my memories of this narrow gateway into the mountains seem to have been ill-starred. In 1984, in equally wet and miserable circumstances, I was settling down for the night when I decided to take a swig from my water bottle. I exploded, my mouth full of paraffin! It transpired that the well-meaning young lad whose job it was to carry my hurricane lamp, fearing spillage on the march, had decanted the oil into what he took for a fuel bottle. Paraffin is nauseatingly persistent in the mouth and all my attempts with toothpaste, lemons and bananas could not relieve me from a night in which I felt more like a ship's engine room than a human being. When next morning, squatting by their fire, I mildly reproached my men with this misadventure, they had rolled about in helpless mirth: it was just the sort of heaven-sent joke they loved. For the rest of the trip, every time they saw me take a drink from my bottle, I had to endure cries of *mafuta ya taa* – lighting oil – and ribald laughter.

There was no point in over-staying at this wretched spot and the following day, leaving the others to the thankless tasks of doing the washing-up and packing the rain-soaked canvas, Christabel and I set off in dense mist up the valley, almost at once crossing the Kanyankoko stream by means of brown, water-rounded stepping stones and entering a wider and more level region. What a transition! Soon after we had set off, the mist had started to shine with silver light and then begun to break up, revealing heart-lifting glimpses of blue sky. Now, with every step, the hitherto dense vegetation opened up into picturesque glades beside the bubbling Bujuku, a stream that would not have been out of place in a Scottish glen. The glades fell between clumps of hypericums and senecios, the latter being the more delicate precursors of the giant groundsels that are such a feature of the higher altitudes. These lower-altitude cousins are tall – 20 to 30 feet – artistically branched, and carry fine, freely flowering spikes of bright yellow daisy-like flowers. Thankfully enjoying more open walking, we found beneath us a springy sward of coarse grasses and alchemilla, the latter a fragrant, silvery-green 'ground-ivy' that is a feature of much of the close ground cover of the sub-alpine zones.

The colouring of the valley vegetation is muted: olive-greens and greys, and autumnal golds and bronzes prevail. This makes any source of bright colour more striking, and the region we had now reached is prolific with the Stairs' ground orchid, whose fine spikes stand 2 or 3 feet tall, a glowing rosy magenta against the grey-green of the alchemilla. These orchids are one of the hallmarks of Rwenzori and a pleasing memorial to Lieutenant Stairs, one of Stanley's most loyal officers, who was the first European to penetrate Rwenzori's fastness. Stairs brought a specimen down from the north-west flanks of the mountain to Emin Pasha, Stanley's unwillingly rescued companion, an Austrian medical man and botanist, who bestowed Stairs' name in acknowledgement. Happily, Emin's name was later given to the sweet, freely flowering violets that we now found at our feet.

Overleaf. *Terrestrial orchids*, Disa stairsii, *amongst silver-leaved alchemilla, with immature lobelias in the background. At Nyamuleju, 10,900 feet.*

It seemed scarcely believable that within half an hour of leaving camp we should find ourselves in bright sunshine in this delectable garden. An hour later we were still there, absorbed with camera and sketch book, while Erinesti and Sirasi diligently collected the more inaccessible specimens for us, the rest of the party having long since passed by. At last we moved on up the open valley, recrossing to the right bank of the Bujuku and entering a new world of giant lobelias and tree heaths. These giant heathers, between 30 and 50 feet in height, had been with us increasingly since before Nyamuleju, but only now did they begin to form the close stands that are such a feature of these mountains. Although this type of vegetation is usually and quite properly called ericaceous forest, these trees are in fact varieties of *Phillipia* rather than *Erica*.

The heath forest clings to the side of the valley, but the widening valley floor, scooped and dammed by glacial action, becomes a circular bog about half a mile in diameter. This terrain is a real test of character: a basin of semi-liquid black peat, crammed with large tussocks of carex sedge 2 or 3 feet high. These wobbly vegetable cylinders, each a foot or 18 inches in diameter with whiskered crowns, are called 'stools' by the Konjo. They do at least make comfortable seats for the exasperated walker, who must choose between the heath forest and the moss-covered rocks of the moraines, and hopping from crown to crown of unstable tussocks, or wading through the black and icy slough of the peat bog in between. We tried it all ways; however, the sunshine persisted and our passage was less unpleasant than I had found it on other occasions, and it was not long before we saw smoke rising against the dark heath forest, showing that our porters had settled in at the Bigo rock cavern.

Bigo is a natural centre in the mountains, a splendid glacial mountain cirque which gives access to three of the most important and exciting of Rwenzori's valleys. To the northwest there is the rugged Mugusu valley, that takes one over the Roccati and Cavalli passes, past Mounts Emin and Gessi, to the remote Lac de la Lune and the Ruanoli valley. To the north-east is the broad elevated Bukurungu valley, between Gessi and the Portal range, which leads to the long Lamia valley and so to the northernmost parts of the range. Westwards lay our proposed route, the upper Bujuku valley, which ends between the highest peaks, Mounts Stanley and Speke, and leads over the Stuhlmann pass to the Luusilibi valley and Zaire. At Bigo one has a satisfactory sense of being at the heart of things. This area also exemplifies the bog-moss-forest ecosystem that lies at the very heart of Rwenzori's role as the most permanent source of the Nile. For this reason I was determined that we should stay here as long as was necessary to do it justice.

With plenty of time and a fine afternoon, we pitched our camp with care on the water-soft ground between the bog and heath-cliffs. There is a down-at-heel Uniport circular aluminium hut here, which is uninviting but good enough to serve as a shelter for cooking, and Ingrid set up our kitchen within. Our domestic arrangements were improving every day. Ingrid is an active Guide leader and I have no doubt that in Britain nine out of ten Guide camps are conducted in cold rain and mud: there could not be a better training for Rwenzori. The amenity of a well-organised kitchen and the convenience of our roomy mess tent provided a standard of comfort to which I was unaccustomed in the mountains, causing me twinges of guilt which I tried to assuage by doing the washing-up. An added attraction at Bigo is bathing in the delectable Bujuku stream; in Rwenzori, in spite of the fact that the mountain is running with water, it can be surprisingly difficult to find a good place for a bathe.

A paraffin crisis had developed: the combined demands of the Primus stove and Tilley lamp had exceeded expectations. A re-rationing party was expected shortly, with a fresh supply, but in the mean time Wiz and Caroline had learnt Konjo techniques with heather and hypericum firewood. Not only did they keep a kitchen fire going on the sward outside our tents, but they applied themselves to pancake-making – a valuable skill which enables you to use flour without an oven. Egged on by the Konjo, they also picked up the knack of roasting groundnuts and making popcorn. These are excellent when hot, make a valuable addition to breakfast muesli, and provide durable rations for nibbling on the march. The easy and nicely mannered relationship of these young people with the Africans of our party – unpatronising and straightforward – was pleasing to see. The disparity between their backgrounds could not have been greater: on the one hand, young women brought up in affluent Britain, with the sophistication of university life; and on the other, the material poverty and grass roots simplicity of Konjo home life. To deny that they were poles apart would be absurdly unrealistic; yet the challenge of life in the mountains and our ungrudging recognition that the African local people were the ones with the skills that mattered, neutralised that polarity and allowed a relaxed and affectionate relationship without condescension on either side. This is something which, even with the best intentions, is difficult to attain in the corrupting outside world, whether on the plains of Africa or in the cities of the West. Success in achieving it can be a refreshing and salutory experience, but you must remove your Western blinkers before you can relax and enjoy it. My young companions did just that. This was humbling to me, because I could remember how long it had taken my young self, in the days of my initiation to Africa, to break down my reserve, and I felt envious of young people born into a world where greater acceptance of cultural differences is beginning to prevail. Now for this brief interlude, time-suspended as we were in our new world above the clouds, for all of us the rules of life were turned back to the innocence of childhood.

Opposite. *Stellate lichen (about ½ inch across).*

Below. *Many varieties of balsams (Impatiens) occur on the mountains: these blossoms, about 1½ inches across, were photographed at 9,000 feet.*

THE ENCHANTED FOREST

Life at Bigo is a continuing ebb and flow between cheerlessness and the sublime. From hour to hour, our spirits reflected this as we swung between despondency and euphoria. The basin lies at 11,300 feet and usually presents a combination of cold mist, saturated moss and black peaty bog. But above the mist the sun is never far away and when it breaks through, the transformation to entrancing fairyland is instant and one's spirits soar.

It is a unique ecosystem, and with our intended prolonged stay, I was able to subject it to close examination. Accompanied only by Erinesti as my camera carrier, I set out to explore and was soon absorbed by a fantastic world that, quite unknown to plains-bound humankind, has its existence high in the clouds above the heart of Africa. The infrastructure of this fantasia is provided by four kinds of tree; the tallest of these are the giant heathers, which grow with contorted artistry to heights of 30 to 50 feet, their trunks, where bare, having a pleasing bronze, squamous bark, while their crowns are made up of filigree leaves indistinguishable from our British heathers, as indeed are the flowers when, rarely, they erupt into massive pink or white blossom. It is these erica and phillipia heaths that dominate and put the characteristic stamp on this zone.

The rapaneas approach the height of the heaths and in favoured niches they may become the predominant species. These can be handsome trees with a rhododendron-like aspect, their thinner branches encrusted with tiny green buds and bright red or purple berries. Hypericums flourish on the drier sites: at their best they can be pleasingly shaped, slender trees, with large, golden-cupped St

John's wort flowers. The last major constituent of the vegetation is a type of giant groundsel, multi-branched and mop-headed, with fine golden spikes of daisy-like flowers.

If I sound slightly reserved about these trees that occur so plentifully throughout this belt, it is because it is not often that one can find an even near-perfect specimen, for they form the aerial support for the other and more dominating life-form – those lowly, elemental plants classed by botanists as bryophytes and pteridophytes, but more commonly known as the mosses and club-mosses, liverworts, ferns and lichens. Not only do these carpet the ground, but they envelop the trees, weighing them down, distorting them and apparently impoverishing them so that most commonly they present a shaggy and tattered appearance. The mosses and lichens are the main contenders for tree space, and the most widespread of all is the 'Spanish moss' – that is, the usnea lichen, which drapes vast areas of the forest with old men's beards of greenish or yellowish-grey fibrils, which contribute so largely to the overall colouring of Rwenzori. But on the stronger branches, the true mosses gain ascendancy, wrapping every limb with billowy muffs, ruffs and cushions in a striking range of colours from brilliant greens, through gold and bronze, to rich Bokhara reddish-browns. These have little of the modesty of our European mosses, but are more like masses of alpine flowers, so generous is their star-like structure and prolific their fruiting bodies. On these aerial garden beds a variety of ferns establish themselves, often hanging as gracious curtains which beautifully transmute the rare sunlight, their silhouetted brown spore cases making striking patterns, while orchids form vein-like meshes of roots.

Two species of giant lobelia – *gibberoa* lower down and *lanuriensis* at higher altitudes – are also striking constituents of this community. When immature they grow at ground level like

Opposite. *Heath-moss forest at 11,000 feet in the Kuruguta valley. The phillipia heath trees are covered with antitrichia moss. There is silvery* Helichrysum stuhlmannii *in the foreground.*

cabbages; in adolescence they are 10-foot-high mop-head palms; then suddenly they send their spectacular flowering spikes 25 feet skywards. Unlike the trees, these herbaceous plants are lightly built and ephemeral, and offer no chance to the otherwise ubiquitous epiphytes.

So much for the world above; as for the ground itself, this is always uneven, for wherever level ground exists bogs form which are inimical to the trees. Beneath the forest there may be peaty beds and tunnels, or the rocky jumble of the moraines: it is scarcely possible to know which, because over past ages the trees have been falling. All these woods are hard, and none harder than the briar trunks, and their innate resistance to rot and biological recycling is enhanced by the prevailing cold and acid conditions – it is a bog oak situation. The result is a wild confusion of tangled dead timber which is completely blanketed by the mosses and ferns, as well as club-mosses, ground orchids and beds of alchemillas. Again, the predominant colours are greens, golds and bronzes; only here and there is some bright contrasting colour provided by the rosy disa orchids, pink or red balsams and purple cardamines – the ladies' smock. A ubiquitous vine, galium, helps to bind all these together.

Constantly subjected to rain, the vegetable sponge I have described is almost always saturated with water: grasp a handful of moss, wring it with your fingers and water will pour forth. There must of course be an ultimate escape for so much water, and as you stumble through this exasperating medium, you come across little rills, sometimes deep down in black clefts, sometimes superficial, that constitute birthplaces of the Nile. The water in such surface streamlets is sweetly pure and translucent, and through its few inches one can see greensward beds of short aquatic grasses, while their banks are clustered with a wonderful variety of shiny dripping liverworts. Surely this must be the world's prime habitat for these usually neglected forms of life, and they throw

up their curious fruiting bodies generously. The sweet water gardens are often banked with cushions of miniature-flowered pearl wort, and myosotis and tussock grasses, while crammed in on every possible surface lichens grow: corals, trumpets, stars and splendid foliate palms larger than a man's hand, all bursting with strange, and indeed sometimes rather repulsive, rufous-coloured reproductive organs. At first, fungi seem strangely lacking, until you realise that in order to survive the cold nights, they have opted for miniaturisation. Then a world of tiny fruiting bodies to be measured in millimetres is revealed.

The club-mosses – lycopodiums – are worth noting. These plants that grow so modestly, an inch or two high on our northern moors, grow here as handsome stout spikes of spiralling interwoven claw-like fingers 2 or 3 feet tall, in prehistoric clusters. Perhaps the moss and fern forest of Rwenzori gives us some idea of the ancient forests that laid down our European coalfields, and we are indeed told that modern lycopodiums are the last surviving relatives of the tropical trees that formed those seams. Their very name, *saururus*, refers to the age of prehistoric reptiles.

To return to the trees with which we started, this whole mossy carpet is a nursery and wherever you look you will find seedling heaths, hypericums, groundsels and rapaneas, all hopefully competing: indeed, these seedlings usually look extremely healthy, for they have not yet been exploited by the epiphytes. In particular, the rapaneas never look so beautiful as when, 3 or 4 feet tall, they present their

---

Opposite. Disa stairsii *and* Satyrium robustum. *These two sturdy terrestrial orchids are found flowering freely in the 9000- to 12,000-feet belt. The disa is shown characteristically amongst alchemilla; it is commonly 2 feet tall, with flowers varying between coral pink and magenta. We found the satyrium growing in boggy conditions at 9,800 feet in the upper Mubuku valley, its flower spikes about 4 feet tall.*

Disa stairsii

Satyrium robustum

X 1/4

X 1/4

unblemished, elegant, glossy leaves to transmit the sunlight as rich scarlets and greens. The lobelias too are profligate in their reproduction, and their infant cabbage-like rosettes are to be found everywhere.

In my description of this strange vegetational complex I am of course only speaking of what one can see: beyond this there must exist an invisible world of microflora and microfauna that completes the well-being of this system, maintains its stability, provides such slow biological recycling as the cold allows, and

mention the aesthetic and almost mystical effect that this environment seems to have on those who dare to intrude. When I first encountered it in 1959, I tried to describe my feelings in my diary, unaware then of the writings of the early Rwenzori explorers: yet, when later I came to read these, I found that the sense of mystical other-worldliness was a common thread.

D. W. Freshfield, who was President of the Alpine Club, was one of the first visitors to the Mubuku valley in 1905 and reached the Moore glacier. He wrote, 'my impressions are amongst the most vivid in a lifetime of travel. That enchanted forest has a weird and grotesque effect that is all its own . . . in rain, a nightmare caused by studying illustrations of Paleolithic vegetation . . . In the rare sunshine one may fancy oneself in the scenery of a Russian pantomime . . . you may be familiar with the Alps and the Caucasus, the Himalaya and the Rockies, but if you have not explored Rwenzori you still have something wonderful to see . . .'

A year or two later, de Fillipi, the Duke of the Abruzzi's amanuensis whose task it was to transcribe the Duke's personal journal, quoted, 'No forest can be grimmer or stranger than this . . . it is primeval, of a period when forms were uncertain and provisory. Silence is profound, absence of any sign of life completes the image of a remote age before the beginning of animal existence, such as might have been those forests which have given us the strata of the coal fossils.'

In 1935, Patrick Synge, a member of the British Museum expedition to the southern valleys, wrote, 'a monstrous and unearthly landscape . . . probably the higher zones of Rwenzori would be voted the most mysterious and unearthly places in the whole of Africa by those few who have visited them.'

When I speak to people of the modern

ensures the constant service that the Rwenzori sponge renders to Africa and the Nile.

Such is my attempt at an objective non-botanist's description of this unusual community that covers immense areas of the range. But only a philistine could leave it at that, and not

Above. *For just a minute or two the sun lit up this world of aerial moss gardens above my tent in the Kuruguta valley at 11,000 feet. The 'mop heads' are immature stems of* Lobelia lanuriensis.

generation who have visited Rwenzori, I find the same sense of wonder at something that is beyond description, but they naturally express themselves in modern parlance. 'Sheer Disneyland fantasia' – 'something from another planet' – 'a world of Arthur Rackham' – 'straight out of Tolkien'. There is a feeling that one is knee-deep in a three dimensional form of impressionist art, verging on the surrealist.

On my Bigo day with Erinesti I did not restrict myself to the heath-moss forest. This sponge system is vital to Rwenzori's function as a mediator of rain, but the carex tussock bogs provide the essential next stage of water catchment, and the Bigo bog is a classic example of

Below. *Erinesti Kitalibara, Christabel King's personal assistant, holding hypericum blossoms.*

this. When the glaciers sculpted the valleys they created a series of cwms or basins of varying depths. These filled with water and, where this was deep enough, they have remained to this day as the enchanting score or more of tarns and small lakes that so perfectly set off the beauty of the mountains. But where the scoops were shallow they have silted up and become colonised, first with golden water moss – sphagnum – and then by the Rwenzori carex sedge. This takes the form of tussocks – each a wild shaving-brush head of spiny culms a foot or two long, often bearing brown and cream flowers. Year after year they build up a matted felt-like base which extends upwards, until the bogs are populated by innumerable flexible but tough tower blocks. I have already described the difficulty of travelling through such bogs but now – with poor Erinesti, his bare legs deep in the cold black mud, to help me with my camera and tripod – I made a serious effort to make a photographic record. My own feet, by the way, in the unlikely event that it is of any interest to the reader, were warm and dry! This was a tribute to the Yeti system of snow and mud gaitors, whereby waterproof, knee-length gaiters are fitted to one's boots by means of a watertight seal. These have greatly ameliorated Rwenzori travel by neutralising one of the mountains' secret weapons – cold wet feet – and I had made sure that all my party were equipped with them.

Erinesti was a good companion. He was not just anxious to please, but he was a quick learner and intelligently anticipated my requirements. He really tried to find out the reasons for what we were doing and sought to understand my fears for the mountains and the loss that may be sustained by his people.

'These very photographs we are taking', I said, as shivering with cold he planted my Benbo tripod in the black glutinous mud, 'and Christabel's paintings, will be printed in a book that will tell the whole world of the beauty of the Konjo's realm, and urge them to come to your aid to protect it.'

'And me – shall I be in the book?' he asked self-consciously.

'Certainly,' I said, 'Such is the help you give Christabel, that we will find a special place for you.'

The carex bogs would present a rather uninteresting scene were their edges not exhibition grounds for the next giant lobelia in the hierarchy – Bequaert's lobelia. Of all the strange plants of Rwenzori, these are perhaps the most bizarre, and they furnish a well-recognised logo for the mountains. Later on I shall discuss the phenomenon of equatorial alpine gigantism, but I cannot leave the Bigo bog without some mention of these plants. Their rosette 'cabbages' are 2 or 3 feet across and a rich purple-green colour. While the plants are immature they squat on the ground; suddenly they shoot upwards, outfolding their leaves origami-style, and then launch skywards a giant monkey-puzzle-patterned cylinder, 6 inches thick, of bracts that contain the deep blue flowers. The final obelisk may be more than 12 feet tall. They tend to occur in family groups and in the mist there is a curious anthropomorphism about them – they suggest extra-terrestrial beings that, trifid-like, will move as soon as one's back is turned.

When Erinesti and I, black with peat mud, returned to camp, in the late afternoon, we found that we had been preceded by one of our Konjo hunting parties and I was called to their camp to see a beautiful blue monkey they had trapped. Blue monkeys are common up to this altitude: they are caught by setting snares in the trees and providing a few poles to lead the monkeys to their fate. The specimen was about 2 feet long from nose to tail-base, with a tail perhaps as long again. Its special beauty lay in its 'blue' coat, thick and silky, made of minutely black-and-white-flecked hairs, and its touching human black-and-white-etched face. Old Kajangwa, our wily trapper, generously offered a haunch for our kitchen but I politely declined – I have never felt I could eat primate flesh. The

Konjo, on the other hand, prize it above other meats. Apart from their flesh, the monkeys are valued for their pelts. The entire carcase is evacuated through an incision in the neck and chest, leaving a skin bag that, when cured by sun and smoke, makes an excellent small rucksack, the skin of the long limbs being adapted to make shoulder and head loops, so that it can be carried Konjo-style.

I left Kajangwa at his work and went to discuss these matters with my companions. To them I had to confess to feelings of guilt and ambivalence about Konjo hunting. My purpose was to urge the conservation of Rwenzori as a national park, and that must include protecting its animals. None the less, even if it had been in my power I was not prepared to forbid our men to hunt. Rwenzori is the Konjo's domain, no one else's. They have been making their hunting trips for countless centuries. In ones and twos, with almost no artefact gear – a knife, some matches, a cooking pot and water gourd, a bag of meal and a sack that serves as a coat and bed – they trap monkeys, hyraxes and duikers with springe snares made of sticks and natural twine. Who amongst us in Europe could approach such skill? What right have I, arriving well fed from Britain, however burning my zeal for wildlife conservation, to object? I do not object and I do not even let the Konjo know that I question it. What I value most is a good relationship with the Konjo, and if we are to ensure their co-operation in the creation of a national park, the maintenance of that relationship will be essential.

It is true that these hunted animals have been considerably reduced in recent years. Such is the extent of the mountain range and the virtual inaccessibility of its remotest fastnesses,

Overleaf. *In the dead-still mist, the weirdly statuesque lances of* Lobelia lanuriensis *and of* L. gibberoa, *20-25 feet tall, are dominated by phillipia heath trees 40 feet tall, draped in usnea lichen. The picture was taken about mid-day above the Bigo bog at 11,300 feet.*

however, that I believe communities of these species will hold out for long enough for hunting to be tactfully phased out under the aegis of a park managed by the Konjo themselves. A system of registration of established hunters, time-restricted licensing, and buying out of rights could be envisaged.

Their most frequent quarry is the hyrax. These are thick-furred, rodent-like animals weighing between 5 and 10 pounds and they have remarkable toes, such that they can run up the vertical surfaces of rocks and trees. It is the anatomy of these toes that gives rise to the much-quoted taxonomist's view that they are related to the elephant. Their beautiful coats are a response to the cold, and they are much prized for making cloaks and hats. They live in dens below the rocks and one seldom sees them in the daytime, but they make the night hideous with their whistling and shrieking, as though the valleys were possessed by fiends.

In days gone by Bigo was a favoured place for the Rwenzori red, or black-fronted, duiker – a small antelope with a bright rufous coat and a black band running up its muzzle. Duikers are the size of a small goat and probably never weigh more than 50 pounds. The moss bog would appear to be quite unsuitable for cloven-hoofed deer, but their secret lies in those very cloven hooves: like those of the larger sitatunga antelope of the lowland papyrus swamps, these are prolonged, providing a 'snow-shoe' facility. Their horns are short and laid back on their heads, giving them a streamlining that enables them to dart nimbly through the tangle.

I had a memorable introduction to these duikers on my first visit to the mountains. Dr Fleetwood, who was then mammologist at the Coryndon Museum, had asked me to obtain a skull and I had mentioned this to Kule. As I was settling into camp at Bigo he called to me urgently that he could hear a duiker – they make a bleat like a modern telephone. Before I could protest, he picked up my gun and led me into the moss forest, accompanied by two dogs and

one of our porters, whom I called Robinson Crusoe, because he was raffishly dressed in a coat of hyrax skins and a monkey-skin hat complete with the monkey's tail bobbing behind. The next hour was agonising: all my motility in the mountains is based on low gear, but here we were racing, ducking, crawling up and downhill, through bog and stream, bleeding, soaking and gasping for breath in the thin air of 11,000 feet. At last I had the duiker in my sights and was about to squeeze the trigger when the mossy matrix collapsed under me, precipitating me into a black underground cavern of dank peat and rushing icy water in which I was immersed up to my waist. The situation seemed desperate – I should be swept along underground to who could say what fate. I clung to a loathsome greasy root, and then, looking up, the hole through which I had fallen was darkened by the peering faces of Kule and Robinson Crusoe. They were speechless with laughter, and only as an afterthought did they hasten to cut a vine and haul me out. By this time we had of course lost all trace of the duiker and returned to camp, my companions bubbling over to tell the rest of our party a well-laced version of my calamity.

On the evening of our last day at Bigo we were joined by a re-rationing party from Ibanda, bringing the worrying news that one of Moses' children was quite seriously ill. Moses and Peter joined us in the mess tent for coffee, and over their exercise book the logistics of our proposed route were re-appraised. It was decided that Moses should go down next day with eight men, to rejoin us later with further supplies via the Mubuku valley. Our loads were already considerably lightened and the remaining twenty-two porters would be sufficient to make the lift up to the head of the Bujuku. Our minds now impatiently anticipating tomorrow's realm of the snowy peaks, we went to bed to the sounds of cascading waters and the shrieking of the hyraxes.

THE ALPS OF AFRICA

To be making one's way day by day up a dramatic valley into the heart of a high mountain wilderness must surely bring a thrill to even the most jaded spirit. Wretched though the rain was next morning, drenching the bramble and briar tangle at the foot of the glacial terrace above our camp, this could not dampen my sense of excitement as we started on the march that was to take us at last to the snowy mountains.

This final stretch of the Bujuku valley, past the narrow second Bigo bog, presents all the discomforts of Rwenzori travel in concentrated form, while at the same time being scenically and botanically hard to beat. A steep final climb brings you to a widening of the valley, where at last the black mirror surface of Lake Bujuku lies before you, with an island rock tower carrying one or two lonely lobelia spikes that shows where the Bujuku stream has its partially subterranean source. Reflected in the lake and rising sharply above it on the far side, above dark green slopes of tree groundsels, are the dazzling ice cliff displays of the Stanley glacier and beyond, the impeccable ice-corniced summits of Mounts Margherita and Alexandra, highest peaks of Rwenzori. So it is true: snow on the Equator – the Alps of Africa!

Well . . . that is how it ought to be, and how I have on just one occasion seen it. Alas, it is rarely so: the most amazing view in Africa is also the shyest, and not one person in a hundred reaching the head of the Bujuku valley has seen it. Nearly always the approach to Lake Bujuku is made in low cloud and drizzling rain, as it was for Christabel and myself on this occasion. Traversing along the edge of the lake, below golden moss-covered cliffs over which ribbons of water cascaded, at last to our relief we saw blue smoke against black rock and knew that we had come to Cooking Pot cave, which was to be our porters' home for the next few days. Ten minutes beyond, a little higher up and at the very head of the Bujuku, in the magnificent cirque between Mounts Stanley and Speke, we came to our campsite.

In 1959 there had been two serviceable corrugated iron huts here. Now, one is a shell lacking both floor and door; the other is doorless, with a badly damaged floor, and is in a generally poor state. However, Ingrid and the girls had already set up our tents, and with the cookers going, we were soon restoring ourselves. I was thankful to be here. My minimum objective with Christabel had been to take her through each of the vegetational zones of the range and spend long enough in each to allow her to secure the paintings we needed. Now here we were, successfully in the last zone – afro-alpine – at 13,200 feet, all of us fit, and adequately provisioned to stay as long as we chose. The prime target of the expedition had been achieved, and anything else would be a bonus.

As we finished our supper and the sun sank towards the mountains, the cloud receded down the valley, leaving us with a crisply clear view of Mount Baker, to the south-east over Lake Bujuku. The rain that we had endured on our way up had fallen as snow on the mountain, and the fearsome north-western cliffs, scene of Shipton and Tilman's classic climb of 1932, were etched in white on black. This was more that Wiz and Caroline could bear and round the fire that night the conversation turned to climbing plans. Next to Mount Stanley, Mount Speke at 16,042 feet is the second highest in the range. It is attractively glaciated and makes a fine ascent from the head of the Bujuku valley, given only – the usual qualification for Rwenzori – that visibility permits. However, I persuaded Ingrid and the girls to make a preliminary trip on the morrow to Lac du Speke, to ensure that they were acclimatised for Mount Speke on the day after.

Lac du Speke lies in Zaire: the names of

places that lie across the international border are French, a relic of the days of Belgian rule in the Congo. To reach Speke's lake you climb up the head of the Bujuku and cross over the 13,700-foot Stuhlmann pass, and then struggle through thick helichrysum scrub over awkward ground, below the western glaciers of Mount Speke. The lake is set in a dramatic cirque of near-vertical cliffs, tucked into the side of the mountain, so steep that only here and there can a giant lobelia or groundsel cling in a heroic posture. I visited the lake in 1984 and was rewarded by an hour of sunshine which turned the waters a limpid grey. From a rock plateau a hundred feet above, I looked down on to a cluster of the gracious 25-foot-tall spikes of Wollaston's lobelias – the ultimate in floral elegance. Numbers of the iridescent green and black, scarlet-tufted malachite sunbirds, with their two sweeping long tail-feathers, were feeding upside down on the flower spikes. This is a mechanism for pollination: what more immaculate conceptual act can there be in the world of nature?

My companions had an equally successful day and on the following morning Ingrid, Caroline and Wiz, with Peter, all equipped for glacier-climbing, set off for the summit of Mount Speke via the Stuhlmann pass. They were entirely successful and returned to camp in high spirits shortly before sunset after a day of sufficiently good weather. They reported that the lower part of the Speke glacier was in a difficult state, lack of snowfall over the preceding weeks having left it as hard, bare ice which was, however, amenable to their crampons.

When I had climbed the mountain with John Newbolt in 1959 we had found plenty of snow all the way up the glacier, but our day was blighted by lack of visibility. So disorientating was this that we first climbed Ensonga (15,961 feet) by mistake – a not unusual Rwenzori

Opposite. *The Alexandra peak (16,703 feet) of Mount Stanley, seen from the summit of Mount Wasumaweso in Zaire.*

situation – and had to retrace our steps to find the true summit. We made a miserable descent in driving hail and thunder, to find the Stuhlmann pass white with snow that had turned the lobelia obelisks into elegant snowmen, on whom the sunbirds were still feeding!

On our present expedition we spent five nights at this camp at the head of the Bujuku valley. It is a bleak place in normal cloud-girt conditions, but we had a greater than usual share of good weather, and when the sun is shining the view down the groundsel- and lobelia-studded slopes to Lake Bujuku, with snow-spangled Mount Baker behind and the ice cliffs of the Stanley glacier high above to the west, never fails to impress. Christabel, dressed for the cold with woolly hat and mitts, set herself up amongst the groundsels to capture this characteristic Rwenzori scene.

A good deal of our time was spent in discussing the practicalities of accommodating visitors in the mountains. Up to the time of independence in 1962, no attempt had been made to open Rwenzori for tourists. For the few visitors that there had been, the simple huts provided by the Mountain Club for its members, together with the rock shelters, had sufficed. In the ensuing quarter of a century, the sickening recession of Uganda into ungovernableness, with the accompanying breakdown in security, transport and the hotel business, has prevented any tourist development. On the contrary, in Rwenzori it has resulted in the never robust mountain huts being vandalised to the point of dereliction. None the less, a surprisingly persistent trickle of visitors has continued and unfortunately even the light pressure of these is proving too much for the ecologically delicate route. Where there is mud and bog, parties spread ever wider either side of the path, destroying vegetation and leaving an ugly, black, churned-up battleground. Around the camps, more and more timber is cut for firewood. The huts themselves have been damaged to a bewildering extent, often, it seems, just to provide fuel. Mountaineers the world over have depressingly depraved personal habits and the areas around the huts are littered with trash, much of which is biologically indestructible. Worse, many of these visitors are all to evidently ignorant in the matter of personal hygiene, and in the cold and inactive environment, where biological recycling is slow, excrement remains disgustingly visible for all too long.

This route represents such a tiny transection of the mainly unvisited bulk of the range that its degradation cannot be regarded as a major ecological calamity, but it is aesthetically offensive. If, in hoped-for peaceful decades to come, Rwenzori is to serve the Konjo as an economically viable national park, much greater numbers of visitors may be expected to come, primarily for aesthetic reasons, and the problem should be tackled earlier rather than later.

In Kampala I had met Kurt Shafer, Program Officer for the USAID Mission to Uganda. He told me that modest US funds might be available for investment in guiding and porterage facilities in the Mubuku valley, as a 'cottage industry'. I mentioned the date we expected to be at Bujuku and suggested he might join us there. Sure enough, on the third day of our stay, in he walked – a battered but determined survivor of the arduous valley trek. The value of our meeting was thus enormously enhanced by the immediacy of our experience of the miseries – and delights – of the road. These concentrated our minds wonderfully and we had little difficulty in agreeing an agenda for an approach to the problem, the first line of which was that nothing should be done to encourage visitors until the infrastructure existed to absorb them without further detriment to the environment. The second line was that as far as possible the natural difficulties of the route should be left much as they are: we were not in the business of making Rwenzori travel easy. But where progressive damage is being done, essentially in the muddy sectors, I suggested that it would be worth while experimenting with some form of locally devised protection,

for example, baulks of eucalyptus timber.

As for the huts, on the one hand we drew up a list of immediate repairs, and on the other, designs for larger and better replacements. The fuel problem can only become worse, and I suggested that it would be worth experimenting with the use of portaged charcoal. But I am afraid the latrine problem, in conditions where the water table is usually at ground level, has left us baffled, except at sites where blasting into moraines might be possible. Finally we agreed that a suspension bridge should be built to replace the one that used to cross the Mubuku below Nyabitaba since, when the river is in spate, people's plans can be aborted at the start. At a certain stage, possibly under the aegis of a national park, the question of providing wardens for the more important sites should be considered: probably only in this way can the ineluctable tendency of humankind to foul its own nest be contained.

The organisation to implement these simple proposals must essentially be provided by the Konjo themselves. Fortunately in the last year or two, the Konjo of Ibanda, who through thick and thin have clung to the protocols laid down by the old Mountain Club in respect of conditions of service and making arrangements for visitors, had decided to create a more formal organisation of their own – Rwenzori Mountaineering Services. Kurt Shafer and I have been involved in advising them and the outcome is that this organisation will act as agent

---

Opposite. Lobelia wollastonii. *The highest altitude giant lobelia,* L. wollastonii, *must rank as one of the most elegant plants in the world. Specimens are often 20 or even 25 feet tall, but Christabel King has chosen a modest plant of only 12 feet so as to enable her to show its detail. This specimen was amongst innumerable plants flowering at the head of the Bujuku valley at about 13,200 feet, and the artist gives us an impression of the dramatic setting amongst flowering tussock grass, giant groundsels, a giant hog's fennel* Peucedanum kerstenii, *and the cliffs of Mount Stanley.*

Lobelia wollastonii

for handling the American aid funds and getting the work done. No-one with experience in Africa will expect these plans to mature either speedily or efficiently, but the Americans have been as good as their word – the funds have been made available, and since returning from the presently described expedition, I have been back to Uganda to see the work started. Experience must make one reserved as to how successful the Konjo will be in this effort to uplift their village economy: but at least we shall have tried, and I believe, along lines that are compatible with the needs of local people and the atmosphere of these wonderful hills.

Any mountaineer, reaching Bujuku, must find his mind turning to Mount Stanley, which is the most complex, highest and difficult part of the range, carrying the most extensive ice and snowfields. I say mind, rather than eyes, because it is more than probable that during the whole of his stay, he will see nothing of this elusive mountain. In order to have any chance of climbing these highest peaks, it is necessary to camp at glacier level, and to facilitate this, thirty-eight years ago in 1951, the indefatigable Mountain Club had constructed two excellent small bivouacs, triangular tent-shaped huts (pig-arks), each to accommodate four people sardine-wise. One of these, at 14,900 feet, is beside the Elena glacier on the south-east of the massif, and the other, at 14,750 feet, is beside the tiny Irene lakes, on the north-east. Each of these can be reached from Bujuku by routes known to the Konjo.

In 1959 John Newbolt and I decided to attempt Alexandra, taking five porters under Kule, with supplies for four nights beside the glacier. We descended to Cooking Pot cave and crossed the evil carex swamp and for an hour climbed up through a jumbled tangle of giant groundsel and moss. Leaving the path that leads over the Scott Elliot pass, we clambered up interminable steep slopes of moss, rocks and mud, through groundsels and St John's wort, into a sombre rock cirque. By a series of

traverses we escaped from this and, moving over bare rocks at the limits of vegetation, came to a bleak glacier-scraped region where Kule pointed out a few artificially placed stones – one of Abruzzi's camps. Beyond this, across glacial rock ribs, with some difficulty in the mist we found the little pig-ark hut, where our Konjo unceremoniously dumped our loads with ejaculations of despair and soon descended thankfully to the comforts of the Cooking Pot, leaving John and me on our own for the next three days.

During the whole of this time, with the exception of just a few minutes, visibility varied between 20 and 50 yards. In anticipation of this problem, we had brought with us a large quantity of tiny red flags, fastened to thin bamboo wands. Each day, cramponing our way up on to the Elena glacier extension of the Stanley ice plateau, we painstakingly made a route by compass bearing, sometimes deflected by crevasses, marking it by flags.

On the third day, disgusted by the unrelenting mist, we decided to settle for an ascent of Moebius. This small summit, 16,134 feet, was sighted from the west by Franz Stuhlmann, a German explorer and naturalist, who in 1891 was only the second European to set foot in the range. He named it after a German mathematician who, as far as is known, had never visited Africa in his life. Moebius rises in the middle of the western edge of the Stanley plateau, and to reach it meant keeping on our present bearing until, at a chosen point, we should switch to a westerly line. But how, in the white-out, were we to judge the turning point? Having probed for crevasses every step of the way, including the spot on which I was now standing, I called back to John to pause while I consulted my compass. He had been keeping up a mild but persistent complaint that I evidently knew little about the use of a compass, and I had deliberately lengthened the run-out of rope between us so that I could shut my ears to this annoyance. Now I had to shout to make him hear, and for some reason in doing so I took a step to one side when, lo! – the world collapsed below me and in a fraction of a second I was precipitated into a crevasse, together with half a ton of ice and snow that rushed and tinkled into the blue-green depths. Providentially, my beautiful ice axe (hand-made in Zermatt in the days when the Perren brothers made them to measure) jammed across my body, leaving me with my cramponed boots thrashing about in icy space and my eyes just above surface level. John, with an exaggerated affectation of tedium, belayed himself with his ice axe and called out, 'What the hell do you think you're playing at?' On the tightened rope I extracted myself shamefacedly and at once surrendered the lead to John, secretly hoping that there were worse crevasses to come.

But even as I was shaking the snow out of my clothes, for the first time in three days a break appeared in the cloud and we could see not one flag but twenty or thirty, stretching across the ice sheet. At the same time, Moebius appeared exactly where it should have been and I snapped it at once with the compass. Then wonderfully, to the north, the immaculate snow and ice summits of Alexandra and Margherita revealed themselves, like fantastic wedding cakes with tier upon tier of cornices. Almost at once the cloud closed in again, but resisting the temptation to change our objective, we continued on course for Moebius, and within half an hour we were on the summit. Climbing Moebius is no great achievement, but we were delighted to have salvaged this out of our three days of wretched endeavour, and as our reward the cloud once more rolled back and we found ourselves looking straight down a horrifying drop into the Belgian Congo, beyond which was a vastly impressive view to the west over the Semliki valley, the blue western escarpments of the Rift valley and the limitless Congo forests. At the foot of the great west wall of Mount Stanley on which we were standing, we could see the tiny Lacs Blanc and Gris, and the larger Lac Vert, all uncompromisingly black. From no other vantage point is the classic tilted

block nature of Rwenzori's structure better demonstrated.

As we retraced our well-flagged steps in the once-more dense mist, I turned my back with immense regret on Margherita and Alexandra: would I ever again have a chance to climb them? But there were quite enough problems in getting off the glacier safely, since the snow had become sloppy and untrustworthy in the higher temperature of the afternoon, and we were thankful to reach our bivouac safely.

Twenty-five years passed and my failure to climb the two highest peaks of Rwenzori still rankled. But on taking retirement in 1984, I set out alone for the mountains, prepared to spend as long as was necessary to attain a variety of objectives. Although I could scarcely hope, alone and at my age, to include an attempt on Margherita, I none the less took my alpine gear.

Fortune smiled on me: by chance I met a young Canadian, Dave Olinkin from Winnipeg – not a climber but keen to have a go, and a few days later, Karl Swanson, a young American from Anchorage, Alaska. Karl was very much a climber and had been teaching at an outward-bound school in Zambia for a year; he was now back-packing through Africa on his way home. Wide though the generation gap was, we decided the chemistry was right, and with seven porters under Peter Bwambale, we set out for the Irene lakes bivouac on the eastern flank of Margherita, ascending steeply through a rocky amphitheatre, with every usual obstacle of tangled groundsel forest, helichrysum, mud and rock. As the vegetation ran out, in the inevitable dense mist we descended a little to a snow-covered rocky shelf where there was a small tarn and a couple of ponds – the Irene

Overleaf. Lac Vert at 13,600 feet on the Zairean side, is constantly a rich green hue, presumably due to aquatic algae. In the foreground are helichrysums, giant groundsels and a massive specimen of Lobelia wollastonii. *These tend to grow to a greater stature on the west of the range.*

lakes. We sent the porters down but kept Peter, as he had been here before and we hoped that he would be able to help us find a route to the east ridge of Margherita.

Over our meal I told my companions how the name Irene came to be given to this place. During the Second World War, curiously, two Poles had discovered this route up Margherita and bivouaced at these miniature lakes. They both had unpronounceable names, but both their wives were Irenes: hence the pleasing name for these pretty tarns, which may be regarded as the highest source of the Nile. Even here, at 14,750 feet, there is still a sparse flora of dwarfed helichrysums and senecios. In the daytime the tarns are fed by sweet ice-fringed glacier melt streams, but these freeze into silence at night.

Morning dawned with only a thin mist, but it took us ages to make our way over slippery ice and moss-covered rocks – a vile combination – and gain the ridge. This had a pleasant alpine character at first, with mixed rock and snow. We moved up easily, unroped: there was even some patchy sun, with impressive views across the Stuhlmann pass to Mount Speke. But after an hour or two, as the icy cloud closed in, we came to a formidable-looking wall of some 300 feet of black rock, which Peter politely but firmly refused to consider. I was relieved: three on the rope would be quite enough, so I lent him my quilted waistcoat to keep him warm and left him gloomily in a bit of a rock shelter to await our return. We three roped up on my red Perlon line, Karl in the lead, myself last, and the stoic Dave, who had never been on a rope before, as middle man. Karl led off and to my alarm had to run out almost the whole length of

line, securing his passage with several wedges and slings, before finding a belay. Dave followed with commendable aplomb, making up in muscle power what he lacked in finesse. When my turn came, the opposite applied: age and altitude can turn your muscles to water, but a wily old fox is a wily old fox, and however desperate my inner feelings I gave little away and collected the slings as I passed. Two more pitches followed which called for every last reserve of my strength. It always amazes me how psychological strength can push back the muscle barrier, making what rationally is impossible just feasible. After a memorable hour, we had overcome the rock wall and resumed the ridge, now formed of ice and snow.

I had read of the celebrated ice rime architecture of Margherita, and seen it distantly twenty-five years earlier from Moebius, but this had not prepared me for the incredible world of ice we now entered. The ridge became an entrancingly beautiful obstacle course of ice-sculptured formations that surpassed the imagination. This is rime ice, not glacial ice: that is, it is formed by the direct freezing of wind-borne moist air, a peculiarity of this equatorial barrier high above the humid African forests. Such ice is hard, and it defies all rules of structure and gravity, reaching out

Below. A glacier melt stream beside my camp at 14,500 feet on Mount Speke. Such streamlets freeze at night, and constitute highest sources of the Nile.

Opposite. At the Irene lakes, at 14,750 feet on the north-eastern flank of Mount Stanley, there is a small tent-shaped bivouac which we used as our base for the ascent of Point Margherita (16,763 feet). Some vegetation persists even at this altitude and the picture, taken at sunset, shows how snow can prevent the groundsel rosettes from closing.

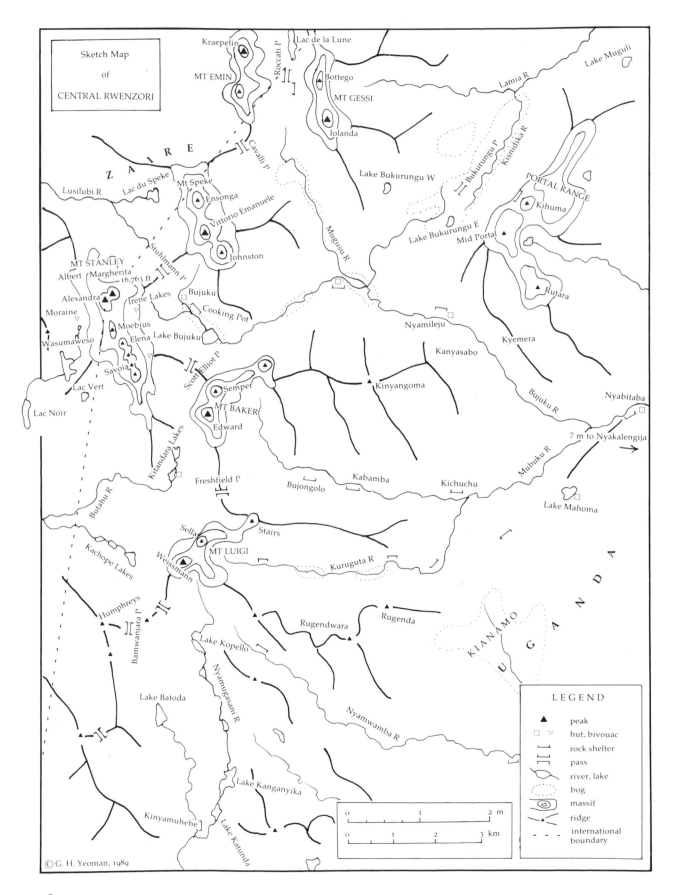

Sketch Map
of
CENTRAL RWENZORI

Kraepelin
Lac de la Lune
Lake Muguli
Roccati P
MT EMIN
Bottego
Lamia R
MT GESSI
Iolanda
Z A I R E
Cavalli P
Bukurungu P
Kisnidika R
PORTAL RANGE
Lusilubi R
Lac du Speke
Mt Speke
Lake Bukurungu W
Kihuma
Ensonga
Vittorio Emanuele
Stuhlmann P
Lake Bukurungu E
Mid Portal
MT STANLEY
Johnston
Mugusu R
Albert Margherita
—16,763 ft
Bujuku
Rutara
Alexandra
Irene Lakes
Cooking Pot
Moraine
Moebius
Lake Bujuku
Nyamileju
Kyemera
Wasumaweso
Elena
Kanyasabo
Savoia
Scott Elliot P
Semper
Kinyangoma
Bujuku R
Nyabitaba
Lac Vert
MT BAKER
Lac Noir
Edward
7 m to Nyakalengija
Kitandara Lakes
Mubuku R
Freshfield P
Bujongolo
Kabamba
Kichuchu
Lake Mahoma
Butahu R
Sella
Stairs
Kachope Lakes
MT LUIGI
Kuruguta R
Weissmann
Humphreys
K I A N A M O
Bamwanjara P
Rugendwara
Rugenda
Lake Kopello
U G A N D A
Lake Batoda
Nyamugasani R
Nyamwamba R
Lake Kanganyika
Kinyamuhehe
Lake Katunda

LEGEND

▲ peak
□ ▽ hut, bivouac
rock shelter
pass
river, lake
bog
massif
ridge
international boundary

0   1   2 m.
0   1   2   3 km

© G. H. Yeoman, 1989

78

horizontally in massive cornices and gargoyles, and vertically in organ-pipe clusters of icicles: there are arches, tunnels, grottoes, stalactites, stalagmites, wind-pruned cauliflowers, filigreed curtains; there are icebergs on a 100-foot scale and gardens of delicate crystalline flowers to be measured in centimetres.

We wended our way, altitude-tired, through this fantastic world, scarcely crediting our temerity. It seemed an act of vandalism to use one's ice axe or to crush the ice flowers below our cramponed feet. And what purity of colour – subtly transmitted blues and greens, spangled with sparkling diamonds. I could not help reflecting on the strange concordance of art and nature that gives Rwenzori the fantasy of the water-saturated moss forest, and then sublimely repeats that fantasy in pure frozen water on its ultimate heights. So this is the virgin purity from whence Egypt's river spills!

Carried away by these flights of fancy and struggling forward with my oxygen-starved thoughts, I was taken by surprise by Karl's American accents.

'Well, I guess this is where the English gentleman takes over.'

I followed his pointing ice axe with my eye.

'Yeah, I guess just another hundred feet to the summit, and you, sir, will please go first.'

Such are the gentle manners of the Alaskans! I didn't argue and within a few minutes we were congratulating each other on the summit which, anomalously in this world of ice, was of barely snow-dusted rock.

Our pleasure at our success and the satisfaction of standing on the third highest point in Africa (how happy I was, I told myself, already to have climbed the other two) were only slightly muted by the deteriorating weather and intransigent mist. All we were vouchsafed were frightening monochrome glimpses of

Opposite. *Rime ice architecture on the summit ridge of Point Margherita at over 16,000 feet. The scale of the curtains of icicles is 10-100 feet. (Photo: Huw Kingston, 1988).*

glaciers falling away to north and south: we saw no sign of Alexandra or Albert, although we knew they were only a few hundred yards away. We spent less than half an hour on the summit and then, bearing in mind the bad reputation of the descent in mist and soft afternoon conditions, made haste to descend. Two abseil pitches brought us to where poor Peter was shivering amongst the rocks, and we safely reached our bivouac with half an hour of light to spare.

One night at this bleak, high-altitude camp had been enough for Peter and in spite of the late hour he at once set off down to Bujuku, leaving us to ourselves. For all their hardiness and skill, so far no cadre of Konjo guides has emerged who are eager to learn serious ice, snow and rock climbing techniques. In the days of the Mountain Club, when locally resident

expatriate alpinists visited the mountains repeatedly and got to know the local men, they were able to take one or two along, equipping and teaching them, and in this way several ascents of the highest peaks were made by Africans. But for many years now, visits by serious alpinists have been irregular: the language barrier is total for most climbers, and equipment exists only on a cast-off basis. Visitors are always short of time and do not want to be delayed by learners. As long as this state of affairs pertains, professional guiding in the alpine sense will make no progress.

There has been a suggestion that one or two Konjo should go to Europe to attend a professional guides course. I think this would be a mistake: they would be introduced to a far more advanced type of climbing than is necessary, and made dependent on too-sophisticated equipment. A more appropriate approach would be for a European or American climbing organisation to send one or two members to the Rwenzori for quite a long period: long enough for them to find out what minimal simple and non-artificial techniques the easiest routes to the snow summits call for, and long enough for them to learn how to overcome the cultural and linguistic obstacles. In this way a cadre of Konjo could develop a genuinely local style of climbing. The existence of such guides could greatly enhance the pleasure of brief visitors who, while sufficiently skilled in their home mountains, badly need local knowledge if they are to travel safely and speedily in this strange cloud-girt world.

In England, before setting out on this trip, I had read de Fillipi's account of the Duke of the Abruzzi's expedition, so over our supper I was able to tell my companions something about the first ascent of Margherita. At dawn on 18 June 1906, from their camp by the Elena glacier, the Duke had set out with his three Courmayeur guides, Joseph Petigax, César Ollier and Joseph Brocherell (the last two having been Sir Halford Mackinder's companions on his first ascent of Mount Kenya in 1899). They

made their way up through 'a gigantic cornice, supported by a colonnade of icicles and aiguilles of ice which at a distance seem like fine white lace work' and attained the summit of Alexandra by *half-past seven* – an incredible rate of ascent, but then all the Duke's itineraries leave you breathless. Margherita was invisible but they made the traverse of the intervening ice and snow col, when 'Petigax cut steps up a steep ice wall . . . and reached the bottom of a cornice where pendent icicles, joining the upright needles, formed a colonnade as thick as trees in a forest . . . supporting the ice dome of the summit . . . and reached a point where a cornice jutted out from the ice wall . . . Ollier served as a ladder for Petigax, who stepped on his shoulders and then on his head with his heavy nailed boots . . . and hauled himself up on his ice axe . . . the ridge was now vanquished and in a few minutes HRH set foot on the highest peak of Rwenzori.' This account of the first ascent of Mount Stanley, transcribed from the Duke's journal by de Fillipi, lacks none of the classic elements we expect from Victorian climbing accounts – *élan*, improvisation and, very literally, class stratification! The Duke named the higher peak (16,763 feet) Margherita, after the wife of King Umberto I of Italy, and its lesser twin (16,703 feet) Alexandra, after the wife of King Edward VII of Britain.

Our tails now up, we descended next day to Bujuku, determined on an attempt on Alexandra, and on the day after followed the route that John Newbolt and I had taken years earlier, up to the bivouac beside the Elena glacier, pitching our igloo tent in a gloomy gully below the ice cliffs. The weather was a little kinder to us than it had been a quarter of a century earlier, but my familiarity with the ice plateau and the staunchness of my friends gave me more confidence. Regardless of the mist we struck out on a north-westerly compass bearing and after an hour or two reached the critical point where we should change direction. Here, as before, the mountain gods relented: for just about fifteen seconds the cloud lifted, showing us Alexandra and Margherita crystal-sharp and excitingly close. I just had time to take the vital bearing on Alexandra before it was once more lost, but this was all we needed. Making our way up a broken ice fall with Karl leading, there was a minor setback when, in a split-second, he disappeared into a crevasse – just the bobble of his Alaskan balaclava remaining visible. Like me, years before, he jammed his ice axe sideways and saved himself.

Now we started ascending more steeply and entered a world of serried ice-rime terraces, like frozen ocean breakers. Systematic route-finding had to be abandoned: it was a case of working our way round or through a series of obstacles of intricate ice sculpture on an arbitrary basis, trusting to our crampons on the good-quality ice. The mist was as dense as ever – we could not see a full rope's length – but as long as we were going uphill we knew that eventually we should reach the summit, and this in due course we did. This summit, at 16,704 feet, is only 60 feet short of Margherita, which remained totally invisible.

Bleak though the summit was and exhausted my state, as on Margherita I could not refrain from philosophising on the extraordinary circumstances in which we found ourselves. The

very existence of this mass of ice and snow on the Equator in the centre of Africa, which was so satisfactorily predicted by the ancients, is a breathtaking incongruity. But when one adds to this its unusual artistry – an art form on a grand scale probably, for its weird style, not equalled anywhere on earth, one may, I hope, be forgiven for asserting that Rwenzori is the prime wonder of Africa and, indeed, one of the wonders of the world. Alas, with current climatic trends, which we are told will melt the very glaciers of the poles, how long will it remain so? What chance has this by comparison insignificant ice cap, against the greenhouse effect?

Rarely have I known such a sense of contentment as when we huddled round the Primus that night. I had confessed to my friends that tomorrow would be my sixty-fourth birthday and they exerted themselves to make this a memorable birthday party. No skill was spared with the rice, bully beef, curry, onions and cabbage. We would not have swapped for the most prestigious restaurant in the world. I was immensely enriched and encouraged by this experience of shared endeavour with these fine young men whose guts, solicitude and sense of humour had enabled me to indulge the wish of half a lifetime.

Next day we parted where the routes divide, Karl and Dave going south to the Scott Elliot pass and the Kitandara lakes, myself retracing my steps northwards to rejoin my men at Bujuku. 'Have a nice day!' Karl called back to me, as their heroic figures disappeared into Rwenzori's helichrysum tangle.

Below. *Christabel King sketching at 13,200 feet at the head of the Bujuku valley below the Stuhlmann pass, amongst giant groundsels and lobelias, on a mat of alchemilla.*

BARE FEET IN THE SNOW

A birthday on the Equator, at over 13,000 feet amongst snowy mountains in the heart of Africa, must surely carry a small element of one-upmanship. My sixty-fourth birthday at Bujuku in 1984 is best passed over: but Wiz's twentieth birthday, which fell after her successful climb of Mount Speke, called for celebration and we did this by lamplight in the hut. A birthday card had appeared; Christabel produced a 2-foot-long salami sausage from a delicatessen at Kew; myself a slab of Kendal mint cake; Ingrid a fruit cake and a candle; and Caroline a rich pudding of pancakes stuffed with apricots and a pair of rather vulgar orange plastic parrot earrings, which Wiz named Emin and Gessi. The porters had presented us with half a roasted hyrax and this gave distinction to our usual hash. I guess this is an anniversary that Wiz will remember long after she has forgotten many others.

Our plan now was to return to our base camp by descending from the head of the Mubuku valley. There is no direct connection between the upper Bujuku and the Mubuku, and one must cross two high passes, the Scott Eliott and the Freshfield, passing through the isolated valley of the two Kitandara lakes in between. In crossing the Scott Elliot pass from north to south one is crossing a high ridge of over 14,000 feet which comprises the col between Mount

Below. *Caroline Massey and Ingrid Pasteur making a pancake lunch amongst the alchemilla at 11,300 feet.*

Stanley and Mount Baker. We were lucky with the weather, but no one should embark on this crossing ill-prepared. In bad conditions it can be a fearsome place, as I had discovered on my first crossing in 1959 with John Newbolt. We had had with us twelve porters under Kulekusengwa, and as we approached the crux of the pass we were savaged by a biting wind and our faces were painfully stung by driving ice grains, while snow lay a foot deep on the ground, concealing the jumbled boulders. Slithering and balancing precariously on their staves in a white horizon-less world, our little column was a pathetic sight. Bare-footed, bare-legged, wrapped in little more than a blanket and pullover, with a heavy load to be steadied with one bare hand and the vital staff in the other, it must have seemed to the porters like a journey to the end of the world.

None the less we slogged on and within an hour or two came to the lower lake where there was an inadequate rock shelter and a small half-collapsed corrugated aluminium hut (since replaced). This contained nothing but a broken cast-iron wood stove which we got going with difficulty, filling the hut with bitter smoke. It was only then, as steaming feet were held out to the fire and several people complained that their toes were hurting, that I realised we had a frostbite problem, and then, as I noticed that their normally smoke-impervious eyes were red and streaming, that we had snow blindness as well. Racking my brains over my frugal medical kit, I came up with a mixture of tincture of iodine and olive oil, with which I set the men rubbing each others' toes to try to induce circulation, while I did the best I could for their eyes with Optrex, for these were the days before steroid anti-inflammatories. We had with us enough rum to lace each man's mug of tea, and in this way slowly nursed our party back from disarray.

On our present journey my companions found this story hard to believe, so sunlit and innocent was the pass. Kitandara provides a lovely lakeside setting for Bequaert's tree hypericum, whose scarlet tulip flowers sometimes lean out like jewels against the black of the waters. (Dr J. Bequaert was a Belgian botanist who made a study of the high-altitude flora in 1915. As we have seen, his name has also been given to one of the giant lobelias.) Also here we found the coloured form of the herbaceous *guilelmii* helichrysum, whose flower clusters, when found before the central blossom starts to fade, are one of the treasures of Rwenzori. The green composite centre merges to brilliant gold at its periphery, where deep rosy-pink bases give rise to the shining silver of the everlasting bracts. The plant grows 2 or 3 feet tall and is remarkable for the intricate web of long silky filaments that enmesh its fleshy stem and leaves, giving it the appearance of being home to an industrious colony of spiders. Presumably this is an ingenious example of insulation against the frosty nights.

On the shore of the lower of these sweetly pretty lakes we came to what is now the least damaged of the mountain huts, a replacement built in 1971 for the miserable tin hut of my first visit. This time we had no frostbite or snow blindness to deal with, but I did find myself facing a larger than usual sick parade. The evening clinic is a daily feature: we were after all totally dependent on manpower – particularly feet and legs, but also the human internal combustion engines driving them. With experience, most of the problems can be predicted. Over the first few days after the porters leave their homes, tummy troubles arise from the change of diet and routine; as altitude begins to have an effect, there are headaches and insomnia, and sometimes relapses to malarial fever; in due course, the wet and cold bring about coughs and sore throats – the demand for cough linctus is well-nigh insatiable. There are muscle aches and painful joints – while sore eyes and toothache may occur at any time. The process is cumulative, with just one or two sick for the first few days, mounting to no less than ten on the present occasion – a 33 per cent casualty rate.

However, the above is only half the problem: the other half is from the knees down. Look at a Konjo's shins and you will see a pattern of shiny scars – and indeed, my shins are in much the same condition. Cuts and abrasions are inescapable in the tangled vegetation, and even slight damage can prove slow healing, leading to sepsis and ulceration if it is not treated properly at the start. As for bare feet, for all their prodigiously tough and horny tissue, they are all too easily penetrated by splinters. Increasingly porters wear some sort of footwear – old rubber trainers or even boots left by visitors – but many prefer to keep such luxuries for smart use in the valleys, while others feel uncertain in them, and prefer their bare feet for security.

African feet have played a large part in my life. When as a young officer I found myself with seventy newly enlisted African askaris under my command in 1942, we were still a barefoot army and nothing was more important than the daily foot inspection. I soon got used to the feel of these almost prehensile, pachydermatous, splay-toed appendages – pale-soled, just as Africans' hands are pale-palmed. Then, and still now, I found something rather touching about them. Before leaving Africa for the war in Burma, we changed, with many doubts and misgivings, to become a booted army. But when we found ourselves in close action against the Japanese, in the monsoon-drenched jungles where mud was our daily lot and silence could mean the difference between life and death, very often we officers reverted to rubbers and the askaris once more to bare feet. Applying a dressing to these foot wounds that will stay on for even one day's march calls for ingenuity born of experience. There is no place here for the individual type of dressing usually provided in first aid kits. Rolls

of 1-inch stretchable adhesive bandage are called for. I took 12 yards on our present trip and used it all.

In these testing conditions the health of the English party remained astonishingly good, our only serious concern being a nasty splinter wound that Ingrid sustained to her leg early on, something which she endured with great stoicism and from which a couple of weeks later she triumphantly extracted a substantial piece of bamboo! We had no problems with altitude sickness (which seems to affect people at much lower levels in Africa than in the Himalaya) – undoubtedly due to our leisurely rate of progress.

The striking difference between the health records of the Africans and Europeans in our party is not difficult to understand. My English companions and I were all clothed and shod in the best modern alpine gear and we slept warm and dry, while on the march we only carried light rucksacks. The bodies of our Konjo porters, on the other hand, were stretched to the limit each day by their heavy loads and in camp, in spite of all their flair for making the rock shelters snug. Apart from the comfort of their fires, they slept cold, hard and often damp.

The pass over which we had just crossed was named after Professor G. F. Scott Elliot, an English naturalist who was the third European to penetrate Rwenzori, after Stairs and Stuhlmann. The name was bestowed by the solicitous Duke, although there is no suggestion that the professor ever got anywhere near it. The crossing of this pass is an important topographical step because one is moving from the eastern watershed to the western. At Kitandara we were at the head of the Butawu, which discharges its water into the Semliki river in Zaire; but it is a secret spot, an isolated Shangri-la, that appears to owe allegiance neither to the east nor the west of the Rwenzori, and because of this it makes a good reflective spot to consider the basic structure of the range.

The underlying rocks of equatorial Africa are ancient granites, and these parts have never

Opposite. *The blood-orange tulip cups of Bequaert's St John's Wort often lean out artistically over open water. This picture was taken beside the lower Kitandara lake at 13,200 feet.*

been subjected to the ocean sedimentation that has altered so much of the present-day surface of the earth elsewhere; thus this region has remained until fairly recent times a weathered continuous rolling plateau. But between 40 and 10 million years ago, the crust of the East African plateau started on a faulting process and the two major features of its geography – the eastern and western Rift valleys – developed. It should not be thought that these developed cataclysmically; rather, their formation was imperceptible, taking place at rates to be measured in a few millimetres per year. In the region of the western Rift valley with which we are concerned, the subsidence was along a roughly north-south line, forming a trench of between about 20 and 50 miles wide – the great trench that contains Lake Tanganyika, Lake Kivu, Lake Edward and Lake Albert. As the earth's crust adjusted to this, some land immediately to the east became elevated in compensation. This is how Rwenzori was created: it is a tilt-block mountain, with a steep face towards the western Rift, and gentler slopes running east. But the Rift is confused and untidy: in fact the once-elevated plain has sunk almost all round Rwenzori, which has been left as a peninsula, only connected to the rest of the old plateau by a neck in the Fort Portal area.

When the block was lifted it naturally carried with it all the material of which it was made, and this included, more or less centrally and running east to west, a belt 2 to 5 miles wide of metamorphosed volcanic rock. This was a stroke of good fortune for mountaineers, because this is the *raison d'être* of the splendid structure of the high massifs, which were in the course of time etched out of the block by erosive forces – rivers and glaciers, the erosion products flowing down and fanning out as silts and slowly filling the sunken areas, a process that was to be compounded later by the products of volcanic activity. Thus while the heights of Rwenzori represent the most ancient bones of Africa, the surrounding Rift floors

represent its most recent skin.

These stress processes are continuing just as rapidly today as ever before, and the earth tremors that are a not-infrequent feature of the area are evidence of this. But comparatively recently, in geological terms, another factor has arisen. The Rifts mark lines of weakness, and it is not surprising that they have become lines of volcanic activity. The country round Rwenzori is pock-marked with low volcanic explosion craters, often spectacularly perfect and, when filled with water and therefore a centre for wildlife, very beautiful. Some of these have been dated as only 5,000 years old, and their effusions, which can be detected as layers in the mountain peats, have created some areas of good soils in the foothills. But it must be understood that, unlike the other high East African mountains that have an afro-alpine flora, which are all old volcanoes, Rwenzori is a non-volcanic exception.

It is a general rule that as you gain in altitude, the mean temperature falls. So it is that as the climber ascends any of these high mountains, he experiences ever-lower temperatures, and these provide one of the most important selection pressures for the vegetation of these equatorial Alps. The way in which plants, through adaptation and natural selection, have responded to the low temperatures is the key to understanding the strange vegetation through which my party and I had been struggling for the last few weeks. Our understanding of this biosystem owes much to the Swedish Professor Olov Hedberg of Uppsala University. As a young plant physiologist he took part in a Swedish expedition to all the afro-alpine areas of eastern Africa in 1948. Many years later my wife and I had the pleasure of welcoming him and his wife Inge, also a professional botanist, to our home in England, and much of what follows is my interpretation, rather crudely simplified, I am afraid, of his views.

This vegetation is profuse in quantity, but in terms of numbers of different species, it is rather impoverished. In its origins it must be descended from the original pre-Rift flora, but the elevation of the tilt-block and in due course the advent of the ice ages must have applied increasing pressures to adapt. Independently the spontaneous rising-up of the great volcanoes to sub-zero-temperature altitudes must have set a similar selection pressure to work, providing a series of genetically isolated biological islands. The consequence is the afro-alpine flora that we have today, some species of which are unique to individual mountains, some to groups of mountains, and some common to all, yet largely found nowhere else.

What are these special environmental conditions? Increasing cold as you go higher is only one. Anyone who has accompanied my party so far will have been left in no doubt about some of the others. The continuous availability of rain, air moisture and ground water is another; but another, in spite of the cloud, is the boundless availability of ultraviolet and infra-red radiation, striking through the thin atmosphere of high altitude from a sun that passes more or less vertically overhead. Surely such availability of energy and water must permit almost limitless plant growth? But such growth can only be attained by means of a highly efficient transport system between the roots and the shoot, which may be growing further and further apart. A rapid and continuous passage of watery sap is needed through the long vascular system, and herein lies the rub. The high-potential system only works for twelve hours a day: for here we are in a region, in Hedberg's words, where it is 'summer every day and winter every night'. The system that works so well in our northern climate – simply shutting down for winter and starting up again in spring – just cannot apply. There is no detectable seasonal variation, only the remorseless daily cycle of high summer and freezing winter, with each transition taking place in a matter of minutes. Some species can take this in their stride: they are simply endowed with a versatile hardiness. For example, on Rwenzori one can find the white-flowered alpine arabis, which is so common in northern European gardens, flourishing at all altitudes. But other species have sought competitive advantage in special adaptations.

In the complex variations of a plant's environment, just one or two critical factors may override all others and determine success or failure. If the obvious factor in Rwenzori is the risk of low temperatures at night, with possible damage to the growing bud, the less obvious anomalous corollary is water shortage: not any lack of soil water, but temporary lack of its availability to the plant because of overnight freezing of the transport system. Sub-zero temperatures may not only freeze the sap in the above-ground parts of vascular plants, but they may drastically reduce the ability of the roots to take up water, and by increasing viscosity they may reduce the rate at which water can move inside the plant. While darkness continues, this may not matter because there is no transpiration demand. But as soon as the sun rises, it makes severe transpirational demands on the leaves, and if the rest of the plant is even temporarily unable to respond to this with a surge of sap, a wilt situation arises. This is the explanation of the anomaly that so many afro-alpine plants living in permanently wet conditions have characteristics that we expect to find in desert habitats: these enable them to shut off demand until their sap system is sufficiently mobilised by rising warmth to sustain it. A corollary of this is that many of the structural adaptations that look so striking are designed to counter the effects of freezing by means of heat conservation and insulation.

The humblest of these adaptations are shown

Overleaf. *The great western wall of Rwenzori, which towers above the Albertine Rift valley in Zaire, is one of the most outstanding sights of Africa. The picture shows its southern half, dominated by Elena (16,300 feet), the Great Tooth, Savoia, Elizabeth and Philip. The view point is at about 13,200 feet on the Kiondo ridge, the highest point reached by Franz Stuhlmann in 1891.*

by cushion plants such as the pearl wort sagina. These pretty clumps grow in waterlogged places, with a mass of minute leaves and pinhead-sized white flowers. They protect themselves by greatly reducing the distance over which sap has to be transported, by forming a dense wad as insulation, and by forming a mass that absorbs energy by day and only slowly releases it by night. The carex clumps and tussock grasses manage in a rather similar way. Others overcome the problem of stem-freezing by dispensing with the stem. These so-called acaulescent species simply lay their leaves as a flat rosette on the ground and lay their flower on that, rather as a water-lily does on the water. The haplocarpha of Rwenzori, with its shiny, heart-shaped leaves and single, central, bright yellow daisy flower, is an appealing example of this. Some thistles behave similarly, making picturesque but prickly mats for the traveller to sit upon.

I have mentioned the critical dawn period when the increasing sunshine may cause water imbalance in the plant. This lies behind the large class of plants that have specially narrow, hard, or thick-skinned leaves, often hair-covered – the sort of devices you would expect to find in desert species. The alchemillas come into this class, to some extent, as do the heathers, but most striking are the everlasting helichrysums, which dominate enormous tracts of the highest vegetational zone, even growing through the snow. Such plants often have silvery leaves, which reduce night-time loss of heat by radiation.

Last of all, but most spectacular, we must take a closer look at the giant rosette plants, the lobelias and senecio groundsels, that have been so frequently mentioned in my narrative, and which have become the hallmark of the afro-alpine zones and compelling subjects for the photographer: Hedberg calls them 'botanical big game'. To take the giant adnivalis groundsel as our model; the immature plant grows like a cabbage at ground level, presenting a handsome leaf rosette of 2 or 3 feet across, made of a mass of broad shiny leaves 18 inches long and 6 inches wide. In the daytime these lie laxly unfolded to the sun, but as the sun sets they close up to form a more or less tightly packed bud which protects the growing shoot. As the tree grows and the cabbage head is elevated, fresh leaves are constantly being opened up in the rosette but the old leaves, dying, remain permanently attached around the stem, so that it is densely wrapped in a thatch of dry brown leaves. This persists for many years, while these long-lived trees approach their full stature of 20 or 30 feet. Eventually they do start to disappear, under the effects of wear and tear, but by now the trunk itself will have developed a rugged protective corky bark. This arrangement acts as a night-storage heater, absorbing heat by day and holding it by night, and also as a duvet, insulating the sap within the stem.

The centre of this stem is a cylinder of large-celled pith which is a reservoir of water that can be drawn on at once at dawn, when the returning sun causes the immediate opening of the rosette and resumption of transpiration. These trees may grow in picturesque small groups or in vast forests that sweep up the valley sides to the snows. It is the existence of these groundsels that makes high-altitude travel possible for the Konjo. Their hollow dead stems provide excellent firewood, while the dry leaves make warm and comfortable bedding and are indispensable for furnishing the rock shelters.

In 1967 Olov Hedberg was able to return to

Opposite. Senecio johnstonii subspecies adnivalis. *The finest of the giant groundsels. Christabel King chose this modest but perfect 15-foot specimen on the slopes above Lake Bujuku, below the towering cliffs of Mount Stanley. The* marcescent *(retained dead leaves) insulation is wearing thin near the ground, revealing a 'corky-bark' replacement. The individual composite flowers are like highly enlarged versions of our common English weed, and attract high-altitude, scarlet-tufted, long-tailed, malachite-green sunbirds as pollinators.*

20 cms

Senecio johnstonii subspecies adnivalis

one of his 1948 study sites on Mount Kenya and identify a photographic frame which he had taken of a groundsel about 5 feet tall. He repeated the photograph and measured the groundsel, deducing that the rate of growth was 1 inch per year. In other words, we must think of the age of mature trees in terms of hundreds of years.

Compared with the slow-growing groundsels, the lobelias, for all their spectacular appearance, are paper-light, quick-growing plants. To take Bequaert's lobelia as a model; these plants stand statuesquely in small family groups – parent, teenager and several babies – connected by underground shoots. The babies, while still squatting on the ground, develop into marvellous circular leaf rosettes, whose whorls of several hundred leaves attain some 3 feet across. The leaves are a dark green colour, subtly transmuted to a deep reddish-purple on the underside, and are covered in fine hairs. Like the groundsels, these rosettes open out generously to the sun in the daytime, but at sunset they close up to make a large protective bud, reminiscent of a gigantic globe artichoke. Within minutes of the sun returning the rosette opens up.

When they are open the central whorls, with their short leaves which protect the germinal bud, are seen to be flooded with water, even

after a dry spell: it has a bitter taste and must at least partly be a secretion of the plant. In the morning the surface of this fluid may be frozen. As the temperature of water decreases towards freezing, there is a 'lag phase' just above zero caused by the actual process of ice formation. This extends the time this reservoir takes to freeze, and the twelve-hour night is never long enough for more than thin ice to form. When the sun rises the rosette immediately re-opens, the ice melts and growth resumes. This system guarantees the germinal bud against freezing.

Below. *An immature rosette of Wollaston's lobelia at 13,000 feet. It is about 5 feet tall at this stage, and closes up at night to protect the growing shoot from freezing. I took this picture by flashlight, after supper.*

An immature rosette of Lobelia bequaertii, *about three feet across, in the open 'day' position.*

Eventually the time comes for the lobelia to raise its fantastic obelisk skywards. As the young shoot rises from the central whorl it is furnished with a wonderful lattice of triangular bracts which provide insulation, while the leaves follow the shoot upwards and continue their protective night closing. The final obelisk may have a leafy stem of some 6 feet, carrying above it a bract tower of another 6 feet – and within each of the honeycomb cells made by the hundreds of bracts is a perfect, deep blue lobelia flower. Such majestic perfection of structure, so superbly adapted to its extraordinarily demanding environment, is totally wondrous. Nothing is neglected in this system, and as the lower leafy stem extends, its wilting lower leaves wrap themselves closely about the lowest part of the stem, insulating it as the groundsels do. The interior of the great column of leaves and flowers is a continuous tube an inch or two in diameter, filled with a milky sap, which is the reservoir that provides the instant surge of fluid at sunrise. Such tubes, with leaves and bracts stripped off, not only provide the Konjo with a wind instrument – a sort of flute – but also serve as water pipes, to tap a supply in a mossy bank.

Bequaert's magnificent lobelia is only one of four towering species found on Rwenzori. At higher altitudes it gives way to Wollaston's lobelia, which is even taller and more spectacular. The bracts of the flower obelisk are in the form of massed silver ostrich plumes. Outlined with a myriad light-refracting drops of mist water as they so often are, they present a picture of immeasurable fineness and elegance, while the actual flowers are more visible than in other species, forming a powder blue cylinder that ascends the spike as flowering proceeds. I have seen specimens of this plant easily 25 feet tall, and they grow into particularly massive specimens on the Zairean slopes.

When I was at school in the 1930s I had a friend with whom I shared a love of the idea of adventure in wild places. I was in a Himalyan phase at the time and could think only of Mallory and Irvine on Everest – but he broke in on my flights of fancy by playing a trump card in the form of an African uncle. (My uncles never seemed to go any further than Eastbourne or the Isle of Wight.) His uncle, so he assured me, had spent his time in the forests and high mountains of the darkest heart of Africa, collecting rare plants and constantly brushing aside the spears and arrows of hostile natives. My friend's name was Wollaston and it was only half a lifetime later that I discovered that his uncle was none other than the celebrated botanist A. F. R. Wollaston, one of the explorers who, in the vintage year 1906, had visited both sides of the Rwenzori, leaving us a lively account in his book *From Ruwenzori to the Congo*. What is more, the spears and arrows were all too real: through what I have no doubt was just another example of the thoughtless mismanagement that characterised so much of early travel in this region, his party was ambushed by temporarily unfriendly natives as they descended the western slopes into the Semlike valley and he was lucky to escape with his life. (No, he hadn't been killed, my friend had said – but there was a trace of regret in his voice, as though his uncle had somehow let the side down in this respect!) Neither of us could then know that this man had the most perfect memorial anyone could hope for, scattered all over the highest slopes of the range, in the form of this most exquisite botanical architecture that bears his name.

There are two other species of giant lobelia on Rwenzori, both occurring at lower altitudes, of which less notice is taken, perhaps because they are found crowded in amongst the dense forest vegetation. *Gibberoa* is the lower-altitude species, and *lanuriensis* the intermediate. They are both outstanding for their height, 20 to 25 feet, and for the elegance of their lance-like structure, which is often gracefully curved to accommodate their habit of growing out from the steep valley walls. *Gibberoa* has unspectacular green flowers, while those of *lanuriensis* are a fine bronze-purple colour. They both have a more lax and open leaf and bract structure, suggesting that at these lower altitudes frost-proofing is not a prime consideration. Similarly, the lower-altitude *ericirosenii* groundsels have a lighter and more open structure.

If high-altitude gigantism is a form of adaptation to the harsh summer-winter regime of the upper mountain, why do we also find giant species – not just the lobelias and groundsels, but also the giant heaths – in zones where frosts can scarcely be a determining factor? Perhaps they are left behind from the ice age? Or it may be that even a slight risk of brief frosts none the less exerts a high degree of selection; this may adequately be met by sheer mass – the night-storage-heater effect. But we should not ignore the positive aspects of the high-altitude equatorial environment. Setting aside night freezing, here we have some of the most intense ultraviolet insolation to be found anywhere in the world, only to be equalled in the equatorial Andes. This is something that the visitor, in these days of awareness of skin cancer, should bear in mind – it is foolish not to wear a hat. This intense energy source is combined with limitless water. Is it any wonder that, in these conditions, plants reach for the sky? Ultraviolet light is an active inducer of gene mutation. May this not be a constituent of the remarkable genetic plasticity that must have lain behind the development of this bizarre flora, that presents us with, in Olov Hedberg's words, 'some of the most fantastic biocoenosis of this planet'. (By biocoenoses he means a special association of plants forming a characteristic and recognisable community.)

I have mentioned the Andes, and on those mountains a fascinating parallel with afro-alpine gigantism is indeed found. Here Espeletia, a genus unrelated to the senecios, has none the less adapted in a strikingly similar manner, while two other species, Lupinus and Puya, have similarly 'copied' the East African lobelias. In other words, in a distant continent, with a quite different original flora, the physiological questions posed by the equatorial high-altitude daily cycle have induced almost identical answers, albeit from a totally disparate gene pool.

Rwenzori, with its equatorial high altitude and central continental position, can be regarded as a uniquely suitable testing ground not only for botanical and genetic, but also meteorological research: not just conventional meteorology as we have come to understand it up to the present time, but the exploding new

Below. *My guide of many mountain trips, Peter Bwambale, sensibly saves his boots from wear and tear by going barefoot. Here, in the Murugusu valley on our way to climb Mount Gessi, he gives scale to a 15-foot obelisk of* Lobelia bequaertii. *They are not long-lived, and paper-light, dead specimens outnumber the living plants.*

sciences that are being forced upon us by the world crisis. By means of such disciplines as environmental chemistry, we need to find answers to questions such as the persistence of radio-activity and world atmospheric pollution, the greenhouse effect and the destruction of the ozone layer. My vision of the future includes a high-altitude laboratory, perhaps a field station of Uganda's Makerere University, working in association with a Western university, and dedicated to addressing such questions. How I envy such scientists of the future!

Kitandara is no more immune to bad weather than anywhere else in the mountains, and yet my memories of it are mostly of sunshine and delectable colours. The greens, bronzes and greys that so characteristically predominate in these mountains can here, in the right conditions, seem less sombre, especially when reflected in the lake. The waters are continuously variable – they do not stay the same for more than a few minutes at a time – while whenever the clouds relent, there are thrilling views of Mount Savoia at the south end of the Stanley massif, the rock towers of Mount Luigi di Savoia to the south, and the glaciers of the western flanks of Mount Baker. In fine weather, which I once experienced, the ridge expedition along the upper edge of the Edward glacier to Baker's summit, Point Edward (15,889 feet), is one of the most delightful in Africa, with dazzling views not only of all the other Rwenzori peaks – it is one of the best vantage points from which to see Mounts Margherita and Alexandra – but also for vast views over the eastern Rift, Lake George and southern Uganda. In foul weather, which I also have experienced, it is execrably vile. On our present expedition, Ingrid, Wiz and Caroline made an ascent with Peter Bwambale: they had a wonderful day, but sadly reported the glacier dangerously depleted of surface snow.

Also from the Freshfield pass one can climb Mount Luigi di Savoia. In 1984 I attempted to climb Point Stairs, an outlier on the north-east

shoulder of Mount Luigi. After a wretched struggle for several hours in dense mist, in a grey monochrome world of bleak, lichen-patterned rock, I settled for an inconclusive summit; my female companions in 1987 had a rather similar experience. Luigi is the southernmost of what are usually described as the six glaciated massifs of the range. Sadly, I must suggest that it is time to give up this pretence. Luigi di Savoia (15,179 feet) no longer carries any glaciation, any more than Mount Gessi does, and probably Emin. It is heartbreaking to report that, wherever one goes, whether on Rwenzori, Mount Kenya or Kilimanjaro, time is clearly running out for Africa's glaciers. I feel like crying out to those who love mountains, 'Hurry, hurry, while stocks last!'

The last night of our five-day stay at Kitandara was memorable. The moon rises late in this deeply sunken valley in the very heart of the range, but even before it had cleared the mountain wall behind our camp it had flooded Mount Stanley, opposite us and across the lake, with brilliant silver light. An illuminated belt slowly moved downwards and as it touched the lake, at about half-past eight, the moon's disc, one day short of full, sailed above the ridge behind us, bathing the scene in a limpid light that we could only marvel at. This is the clarity that the astronauts know. Usually a brilliant moon abates the stars in her quarter of the sky, because of the atmospheric haze: at our altitude we had no such haze and the dazzling disc was supported by innumerable clusters of stars in an extraordinary demonstration of the power and the glory of our galaxy. Every leaf and flower about us; the reflection of the flood-lit mountains in the lake; the cliffs, waterfalls and glaciers, all were sparkling crisply with the ethereal silver light. The shiny undersides of the leaves of the night-closed groundsels reflected the light like silver foil, and the white flowers of the everlastings shone like stars at our feet. We could hardly take it all in, and sat by the fire until after midnight, talking in hushed tones, tired and yet loath to spoil the magic of the moment. But in the end we had to surrender it and go to our moon-illuminated tents, for on the morrow we had to cross the Freshfield pass and start our journey home.

Below. Helichrysum stuhlmannii – *an everlasting or* immortelle – *is one of the most successful plants of the range, being found from 10,000 feet to over 15,000 feet, often as the majority species covering immense tracts of mountainside. It shows great plasticity of form, varying from impenetrable 8 foot high woody scrub to such tiny specimens as these, peeping through a few inches of snow.*

THE KINGDOM IN THE MOUNTAINS

To crawl out of one's tent at dawn and look straight at a massive, nearly vertical mountain wall – only a few yards away – covered in dense wet vegetation, and to know you have to climb it after breakfast, is daunting; it brought us down to earth after our moon-gazing.

The Mubuku valley is more difficult than the Bujuku, although more dramatically beautiful, and it is better descended than ascended. To reach its head one must climb some 600 feet very steeply from Lake Kitandara, and then another 400 feet more gently to the Freshfield pass. This is an open airy place where, for a little while, you can enjoy fell-walking reminis-

cent of Cumbria, in a world carpeted with bronze moss and dwarf silver everlastings. Descending by a series of steep and awkward glacial terraces, you pass the Duke's old base campsite under the great overhanging cliffs of Bujongolo and come to the pretty water cascade of Kabamba, a truly perfect high fountain of the Nile. A little beyond this is a good rock shelter and we were able to pitch our tents for a couple of nights on a dry glacial drumlin. From here, with the porters scenting the comforts of home, we passed the cliff overhang of Kichuchu and went straight on to our old campsite on the ridge at Nyabitaba, descending the following

Below. *Our celebration party on our return to Ibanda. Ingrid Pasteur stands next to Moses Matte (left) and me. In the foreground Peter Bwambale is sitting in front of Wiz. (Photograph: Caroline Massey).*

day to Nyakalengija where we re-occupied our old base camp. Christabel had a good deal of finishing work to do on her folio; I had to collect cash from Kasese and pay off our men; and we had various social duties to attend to.

Of the latter, the most important was a Sunday lunch party that was declared in honour of our ladies, to be held at the local brew house not far below our camp. It was a bamboo, mud and thatch, single-roomed dark building, with a bamboo-fenced compound at the back: men in the house, women in the yard – but the rules were of course waived for my companions. In inviting us to the party Moses had said that they hoped to be able to kill a fowl for the meal. This didn't sound very adequate to me, since the guests were to include all thirty of our porters as well as a number of hangers-on, including the local sub-chief and the village pastor. So when he added that it was a pity they couldn't afford a goat, I took the hint and asked him to find the fattest one he could at my expense. It turned out to be a rather pathetic creature which spent the day before the party tethered in our camp; the tender hearts of our young women found this hard to bear.

Lunch could not start until everyone had returned from church. In the mountains, where the environment so strongly has the upper hand, one could say that a natural animism prevails – a feeling that places have characteristics and indeed spirits that can affect and even control the soul of man and his well-being. But in the valleys and on the plains the Konjo are good Christians and Sunday morning church-going is the social high point of the week. Of the many sad changes that have marked Ugan-

da over the past quarter-century, one is the almost total disappearance of attractive clothing. My early memories of the country are of the most gracefully and brilliantly clothed people in the world – the men almost uniformly wearing decorous flowing white gowns, the women spectacularly coloured and elaborately fashioned dresses. Now everything is drab – except on Sunday morning. Then, as the little churches open their doors at the end of service, there is a burst of colour like a flower festival as the brightly clad women and children pour forth, and even the men who, alas, have largely opted for European style, appear at their smartest.

We crowded into the dark hut for our meal, twenty or thirty people pressed closely together, squatting on the earth floor round large plaited trays piled with *muhogo* dough (it is like plasticine and sits cleanly on the basket-work), and enamel dishes of steamed banana *matoke*, rice and stewed goat's flesh, I found it heavy going. Although I had, as usual, lost about a stone in weight on our trip, this did not give me a sharp enough appetite to make much of an impression on the weighty portion that was pressed upon me. The real business, however, was drinking. There was an apparently limitless supply of *tonto* – banana beer – and *waragi* – the distilled banana spirit that is boasted as Uganda's national drink. *Tonto* can vary from a sweet, almost honey-like and scarcely fermented banana juice, to a fairly potent kind of cider. *Waragi* is more dangerous, and my companions and I steered a delicate passage between not poisoning ourselves and not giving offence to our hosts. Speeches were called for and when my turn came I put together a few words about the beauty of the Konjo heritage, how Christabel's paintings were going to reveal this to the world, and how they themselves must strive to protect it. I said everything three times, once in classical Swahili, one in a debased up-country dialect, and once in English. In each case, two or three happily inebriated interpreters simultaneously

Opposite. *The Victorian geographers were for ever seeking the 'coy fountains of the Nile.' How thrilled they would have been by this perfect cascade at Kabamba in the upper Mubuku valley. Above is heath forest, with silvery helichrysum, young mop head lobelias, and the* erici-rosenii *giant groundsel in the foreground. Extreme left – an immature rapanea (rhododendron-like) tree.*

transposed my words into their imaginative versions in Rukonjo for the benefit of my less sophisticated listeners. At last, after photo calls, we were allowed to leave and make our way back to camp.

Next morning we struck our tents, loaded our lorry in drizzling rain and with several halts for friendly and tearful embraces, left the little valley community behind us and re-entered the Africa of the plains.

I had promised my companions a traditional East African safari and we spent the next two weeks visiting the savannah game country around Lake Edward, the mountain lake district of Kigezi in the extreme south-west of Uganda, and finally making an ascent of Mount Sabyinyo, the 12,000-foot volcano in the Virun-

gas, which shares its summit between Uganda, Rwanda and Zaire. My companions' time had now expired. Our group had been brought together largely by chance, but it could hardly have been more successful. How grateful I was for such lively, good-humoured and intelligent company; for the sheer uncomplaining guts with which they had all borne the discomforts of our unusual journey; and for the unfailingly equable manner with which they had put up with my testy idiosyncrasies! I finally left them all in the protection of a Catholic nunnery at Mbarara, from where, not without problems, they eventually reached England safely.

A few days later I was back in the valley at Ibanda, planning a new journey with Moses

Below. *At the celebration party, decorum required Konjo ladies to keep to the outside compound of the brew-house. It is Sunday after church, and our hostess is wearing a beautiful* **kanga** *shawl over her dress.*

and Peter. In 1984, civil disturbance and army activity in the southern populated areas had prevented my journey up the Nyamugasani valley, and I had had to settle for the more easterly Nyamwamba. We now planned my original proposition in reverse: to go in up the Kuruguta valley and come out by the Nyamugasani. In view of the possible political sensitivity of this project, Old John, Moses and I put our proposal to the District Executive Secretary at the political headquarters at Kasese. He seemed bewildered that anyone should want to do such a thing, but wrote me out an authority without demur.

But on the eve of our departure, 1 September, Moses was taken ill with dysentery and I had to replace him with Elia, a rather ineffectual old chap who did not inspire a great deal of confidence. Worse was to come; when we were only three or four days on our way, in the lower Kuruguta valley, Peter Bwambale was stricken with toothache and I had to send him down to Kilembe where it was said there was a person who could draw teeth.

In spite of these setbacks, we slowly made a route up the wild and unvisited Kuruguta, staying two nights at each camp and cutting the route ahead on the day in between. The rock walls are awesomely steep below the towering heights of Mount Rukenga, and the bogs are intractable, but the rock shelters in the lower part of the valley are good. By 8 September we three elders – Kajangwa the tracker, Elia and I – were at the head of the Kuruguta, reconnoitring a route out to the south-west, when a breathless runner caught us up. He had left Ibanda only the day before, and from his blanket he produced a sodden brown envelope. Opening it delicately, I saw that it was from Old John Matte. 'Dear Mister Yeoman sir, ' I read, 'I must tell you that war has broken out . . .' There were no hard facts: just that it would be folly to pursue our intended route. Whichever side we bumped into, it would assume that we were partisan to the other. The only safe course was for us to retrace our steps to Ibanda. There was

nothing else for it: we returned to camp and the next day moved down to the rock cliff of Kichuchu at the junction of the Kuruguta with the Mubuku valley.

I was just making tea at my tent under the rock overhang, a few yards from where the porters were sitting round their fire roasting groundnuts: it was about four o'clock on a sunny afternoon with plenty of blue sky, when the valley and hills were rocked by two massive explosions, a second or two apart. Looking up I could see no sign of anything untoward. But instantly, in my heart and spine, there began to tingle nerves that had lain quiescent for over forty years since we had faced the Japanese in the not-dissimilar Myittha gorge between the Kabaw valley and the Chindwin river in Burma. The cheerful voices of my men ceased instantly: there was not a word from the figures by the fire.

I felt it was important not to show that I was alarmed, so I completed my tea-making and enjoyed my tea as best I could. But within myself I was in tears of dismay, that beautiful Rwenzori, the quintessence of all that was pure and inviolate, should be desecrated in this way. When at last I spoke to the men, I made as light of it as I could.

'Ah, these idiots. Who do they think they are, disturbing me at my tea? Is nothing sacred these days?'

'Truly, *Mzee*, there's nothing these oafs in the army won't do. They are like children – just because they have bombs, they feel they must let them off.'

Later that evening I joined them by their fire. All spontaneity had gone from their conversation, but haphazardly, with many a grunt and cough, with sudden profound affirmations and denials, we put together all we knew that could lie behind these blasts from a clear sky. I found that I knew things that they did not, and they, of course, knew a great deal that was new to me, and I was much helped by Erinesti's thoughtful mind. Both they and I knew, in our different ways, that these explosions were just one more incident in a sad history of expediency, ignorance and pragmatism on the part of their erstwhile overlords, the British, which had left the Konjo in a state of permanent smouldering resentment against their Ugandan neighbours. Between us, in the pitch dark, as the firelight flickered on the vast rock face above us at 10,000 feet in the heart of Africa, we pieced together the following story.

In the year 1891 the scramble for Africa was at its height. In eastern Africa, the contenders were Britain and Germany, both of them under pressure from the Free State of the Congo – the private empire of King Albert of the Belgians, which was shortly to become the Belgian Congo. When in that year Captain Lugard's expedition reached the land we now call Uganda, his purpose was to establish a British protected block that would offset encroaching German and Belgian claims. Of the four self-styled kingdoms that he found, Toro in the south-west, the weakest and least authentic, had been invaded by the armies of King Kabarega of powerful Bunyoro to the north, from whom it had seceded sixty years earlier. Kasagama, the young King of Toro, had fled and found safe haven with his Konjo neighbours, who held their natural domain around the mountains. He was in due course smuggled to Lugard's camp in the east of the country. Lugard and his tiny force were hard-pressed, his only support coming from the kingdom of Baganda – the dominant but uncertainly loyal tribe who lived in the region around the north-west of Lake Victoria – and he saw in the Batoro potential allies against Kabarega. He marched westwards across Uganda, liberated Toro and re-instated Kasagama on his throne, extracting from him in return a treaty of protection. In drawing up the treaty, Kasagama was given a free hand in delineating his kingdom, and disregarding any debt of honour that he might have been expected to acknowledge to the Konjo who had saved him, he included within his boundaries the Rwenzori

mountains and the adjacent south-eastern region of foothills and plains, called Busongora, that was the Konjo heartland. No opinions were expressed about this by the Konjo, because they had not the slightest idea of what was going on.

At this time, the Konjo tribal land extended both east and west of Rwenzori. But in 1910, again without any reference to the Konjo, the Anglo-Belgian Boundary Agreement drew the international border down the highest spine of the range, dividing the Konjo ethnic region in two. Thus one injustice was heaped upon another.

For the time being this injustice had little practical importance to the Konjo. But in the steady process of social and economic development that ensued under British protection, Toro adopted a racially biased policy that discriminated against the Konjo in politics, education, social services, language and taxation. Rising Konjo resentment led to protest and in 1921 three Konjo leaders – Tibamwenda, Nyamutswa and Kapoli – were captured and hanged by the Toro with British acquiescence. These grievous affronts might have been borne by the Konjo, who are stoic and self-sufficient, but the psychological impediments and verbal abuse that so curiously accompanied them were more than they could bear. They were compelled to refer to themselves as Toro and their own language was ignored to the point of suppression. No reference to their existence was allowed in the constitution, although they comprised one-third of the population of the Toro district. They were treated with arrogance and commonly referred to as monkeys, insects, dogs or pigs.

The smouldering Konjo resentment began to

Opposite. Hypericum bequaertii. *The brilliant blood-orange tulip-like cups of the high altitude St John's wort contrast dramatically with the low-key background hues of the sub-alpine belt. These flowers came from graciously shaped 40-50 feet tall trees beside Lake Kitandara at 13,100 feet.*

Hypericum bequaertii

glow and was fanned into a small flame. In 1954 a certain S. R. Bukombi started a Bakonjo Life History Research Society. In origin it was a non-political focus for Konjo culture and history. However, in this same year, a young Englishman, Tom Stacey (who was in the course of time to become a Member of Parliament), was back-packing about Africa. He embarked on a long journey through the inhabited areas of Rwenzori, having picked up a young Konjo companion, Isaya Mukirane – an unemployed and disaffected primary school teacher, a victim of Toro prejudice. As their journey proceeded, Tom Stacey's company and influence had a catalytic effect on Isaya Mukirane, which led to him taking over the Bukonjo Society and giving it the orientation of a liberation movement.

By 1961 the Konjo at last began to realise that the British were serious in their proposal for independence for Uganda, and that this could be the time to throw off the Toro yoke. A new electoral system was being introduced which for the first time gave the Konjo representation in the *rukurato* or district council, and Isaya Mukirane was elected. When he led a protest walk-out, he was arrested but released on bail. He broke bail, returned to his native hills, and deep in the forest of the southern part of the range close to the Zairean border, raised his standard of rebellion, announced the Rwenzururu Resistance Movement and the birth of the Independent State of Rwenzururu: all this, a few months *before* Uganda itself became an independent country.

A Central Government Commission of Enquiry produced a statesmanlike report which upheld all the Konjo grievances and condemned the Toro Government for the obtuse and insensitive manner in which it had handled the matter. But its sensible recommendations fell short of the creation of a separate district, and for the Konjo it was too little too late. Their sights were now set on total separation not only from Toro, but from Uganda.

A reign of terror on both sides ensued, with raid and counter-raid, and wholesale imprisonment of Konjo by the Toro authorities. In February 1963 a state of emergency was declared, and the Prime Minister Designate, Milton Obote, ordered the still-British-officered Uganda Rifles to seal off the mountains. But Isaya Mukirane had not gone alone into his mountain redoubt: as well as thousands of irreconcilable spearmen, he had taken with him those most potent of weapons, a sense of outraged national pride, and a crucible of national heritage and culture, and these were to prove enduring. Obote and his new Government sympathised with the Konjo and hesitated to order the army into the hills. For a moment, everything hung fire.

And now a quite extraordinary change of scenario takes place. The driving icy rain is no longer that of high Rwenzori, but that of Northumberland Avenue, just off Trafalgar Square in London. The time is January 1963: people are leaving after a lunchtime lecture at the Royal Commonwealth Society, and an African is about to get into an official car. But as he does so, he recognises an Englishman standing on the pavement. The Englishman is Tom Stacey, now an overseas correspondent of *The Sunday Times*. The African is Timothy Bazarrabusa, the only Konjo at the time to have attained high office of State: he was Uganda's first High Commissioner in London. He had been a chairman of the Bukonjo Society and in due course was to become the first African president of the Uganda Mountain Club. To Stacey he said,

'You know what has happened to our poor Konjo?'

'No.'

'They have rebelled against the Government . . . They took to their spears last September. Now they have got everyone against them – the Government of Toro and the Central Government. Nobody can do anything with them.'

He got into his car and drove off, but later an invitation to lunch revealed the whole story to

Stacey. Obote had reasonable compromise proposals to convey to Isaya Mukirane in the mountains but there was no one in Uganda who could take them to him and expect to be received with anything but hostility. Over the fruit salad, as Stacey puts it, Bazarrabusa asked him if he would undertake this difficult, arduous and possibly dangerous assignment.

Stacey tells us what ensued in his book *Summons to Rwenzori*. He has a good story to tell and he tells it well. Armed with his message from the Government and a guarantee of safe passage for Isaya Mukirane, he was to persuade him to descend to a lowland meeting place where the resolution of the impasse would be worked out. His journey took him from the southern lowlands of Bwera, where the Uganda Rifles were in control, through what he calls the Bukonjo heartland, up a ridge along the east of the Tako river which forms the border with Zaire, to Isaya Mukirane's mountain retreat of Kitote. When at length he managed to meet him, he found his friend of nine years ago changed: indeed the pathos of the story lies in the transition of his outgoing and delightful young companion of yesteryear into an embittered victim of paranoia verging on madness. Stacey's attempts at shuttle diplomacy, with cleft-stick messages between the mountain, the plains and Kampala, came to nothing. The only outcome was an agreement between the leaders of the moderate lowland Konjo and the Government that left Isaya Mukirane and his irredentist hillmen incensed, with consequential bitter inter-Konjo strife and killings.

Stacey had found Isaya Mukirane a self-styled president of the independent mountain state of Rwenzururu. He had a group of ministers, a Secretary of State (with a typewriter and official rubber stamp), a blue, green and yellow flag, an army of a thousand or more spearmen, a police force and judiciary, numerous primary schools and a teachers' training college where the medium was Lunyarwenzururu – the true ethnic Rukonjo language, purified of Lutoro corruptions. The Secretary of

State was hard at work sending out letters to such addresses as U. Thant, then Secretary General of the United Nations Organisation, the Organisation of African Unity, the Prime Minister of Uganda (as a foreign country), the Pope, the Queen of England and many other dignitaries.

From Stacey's description it would be easy to dismiss all this as the fantasy of a Gilbert and Sullivan opera, such as could only be staged in Africa. The hierarchy that was taking itself so seriously consisted of illiterate barefoot woodsmen dressed in rags and tatters, who lacked both firearms – or indeed any weapons that they had not made for themselves – and any building that was more permanent than the bamboo and thatch of which it was made. Can we, or the Government of Uganda, take seriously this 'village Hampden that with dauntless breast, the little tyrant of his fields withstood'?

Thomas Gray's long-buried hero, in his 'Elegy in a Country Churchyard', was described as 'some Cromwell, guiltless of his country's blood'. Alas, we cannot say that of Isaya Mukirane, but equally, can we withhold from him Gray's admiration? Rwenzururu has, at the time of my writing this in 1989, survived for more than twenty-five years and there is no obvious end in sight. Every government Uganda has had during this turbulent quarter of a century has made one or more serious attempts to end the insurrection, usually starting with negotiation and ending with force, but has got nowhere.

Following Stacey's attempt, in 1964 a large body of Toro spearmen trapped and massacred many hundreds of Konjo, leading to bitter Konjo retaliation. In 1965 Obote once more attempted appeasement, including an amnesty, causing a split in the Konjo ranks that led to murderous inter-Konjo conflict. In 1966 Isaya Mukirane, the humble and despised primary school teacher who was, he fervently believed, called by God to high office, died of natural causes. He had declared himself *Iremangoma* –

'he who rules the throne' – that is to say, king, and his 9-year-old son, Charles Wesley Kisembo, was declared king under a regent.

The irredentists of the new kingdom in the mountains were in the ideal guerrilla situation. They could withdraw into the freezing heights; slip into neighbouring Zaire; or simply merge with the low-level Konjo population. But woe betide such lowlanders if they denied aid to Rwenzururu: they were liable to ruthless attack on lives and property. It is horrifying to think of the gentle Konjo of my expeditions becoming killers: but I have no difficulty in believing it. They would be a terrifying enemy to have to face in their mountain fastness. The thought of the savagery in the mists, the golden moss and young lobelias splashed with blood, and the crumpled bodies lying in the tangled forest tears at the heart. Many hundreds – probably thousands – have been killed, wounded, tortured; houses burnt, lives disrupted. There is nothing Gilbertian about this. I have memories of war in the forests of Burma more than forty years earlier: the thought of it in Rwenzori leaves me sick at heart.

In 1967 a new republican constitution swept aside the anachronistic kingdoms of Uganda. Toro was no more and the Konjo rejoiced. But it made no difference to the hardcore mountaineers, who continued to regard with disdain the goings-on in their foreign neighbour, Uganda. In 1971 Uganda entered its dark age under the rule of Idi Amin who, in fact, granted the Konjo the district status that had been denied them for the past ten years. A new District Headquarters at Kasese meant that at last Fort Portal faded from the scene, but this had no effect on the kingdom in the hills. When at last the young Charles Wesley succumbed to the promise of a university education in America (where he still is at the time of writing), he was simply succeeded by the *Kinyamusitu* – king of the woods – Richard Kule. When Obote returned in 1979 there was resurgence of Rwenzururu activity against the Central Government – what became known as 'liberating

Kasese'.

In 1980, for the first time, there was a gesture of statesmanship. Obote appointed an Mkonjo District Administrator at Kasese, Blasio Maate. He was a Konjo loyalist who had been a founder member of Rwenzururu and had endured imprisonment and exile in the 1960s. He was as near as anyone could get to being *persona grata* with the irredentists. He persuaded them to form a joint committee with the lowlanders, and replaced the government police with Rwenzururu askaris. Negotiations with Central Government were started through the medium of a Konjo elected Member of Parliament, Amon Bazira, and in 1982 it was announced that the Rwenzururu had 'come down from the mountain' and that the independence movement had officially come to an end on 15 August 1982.

Unhappily, far from this being so, poor Uganda was in process of descending into yet another agony of internecine war as Museveni, who was destined to become president, launched his National Resistance Movement. Amon Bazira was incarcerated as a member of the hated Obote's Government and silence fell over Rwenzori. But in the rock shelters of the high valleys the woodsmoke of the irreconcilables still rose blue against the dark cliffs in the never-ending mists.

In 1984, on my return to Rwenzori after an absence of twenty-five years, my proposal to my Ibanda men that our end-to-end traverse of the range should start in the south at Kyarumba and follow the Nyamugasani valley northwards had been vetoed. 'There are soldiers all over the place,' they said. 'It would be more than our lives are worth.' So I had to settle for the more easterly Nyamwamba valley. This was what had led me now, in 1987, to try to complete this southern sector in reverse, through the Kuruguta: and this was what had led us to this desolate evening round the camp-fire at the foot of the great cliff at Kichuchu, our spirits still subdued by the two devastating explosions that had come out of a clear blue sky.

We squatted round the heath-log fire, united in our unhappiness. The long heavy trunks are laid parallel to each other, burning in their centres, so that they extend to right and left, dividing the squatters into two facing groups, like a parliament. From a little way off, the picture was dramatic: the vast, twisted-veined rock-face as backdrop was fleetingly illuminated by oblique light from the fire, casting its rugosities into shadowy relief and causing strange surrealist shapes. Silhouetted against this was the conspiratorial, blanket-shrouded group of figures. It was like something from a neolithic age.

It was agreed that the sounds had come from the direction of Ibanda and that we must assume that our home valley had been occupied by the National Resistance Army (NRA) in belligerent mood. Our return from a southerly journey could only cause us to be suspected of having dealings with the enemy. I kept my counsel, hearing each man out, but trying to rehearse a practical course in my mind. Should we stay where we were and await the turn of events? Or seek another route, perhaps my old northern one to Bwamba? Or even cross into Zaire? No one was happy with any of these, because all the men's thoughts were with their homes at Ibanda.

'Then I suggest we send a scout – someone who can reach the valley by a secret route, avoiding the soldiers' posts. He should take a letter from me and discreetly pass it to *Mzee* John, and bring his reply back to us. Then we shall at least know what the situation is, and can decide on the best course.'

A murmur of approval greeted this suggestion. But who should go? Who was to undertake this potentially risky task? There was a unanimous answer.

'Muthundya – it must be Muthundya.'

'Which of you is Muthundya?' I asked. They all pointed to a slender youth, who was squatting diffidently at the back of the group. The firelight showed a pale brown, rounded face as yet unblemished by age or beard.

'But why should we choose Muthundya, when he is clearly so young?'

'Because he is faster than any of us. He speeds through the thickest tangle like the red antelope.'

Muthundya remained silent while he was being discussed, as if he had neither ears nor opinions.

'OK – but we can't send him by himself,' I said. 'No one should travel alone where there is danger. If there is a second man, he may escape

Below. *Mid-day on the Equator! We are lost at 12,000 feet. Our path-finders out ahead, send back whistle signals only indicating doubts and misgivings. Meanwhile we take consolation from an impromptu fire and debate the merits of brewing tea.*

back to warn us: who else will go?'

Again the answer was unanimous.

'Bukombi,' they cried, 'Bukombi – only he will keep pace with Muthundya.' Like Muthundya, young Bukombi bore this dubious distinction with modesty, no one having bothered to ask him his views on the matter.

'Good – but listen. How many days will they take?'

'Days!' There was astonishment at my question. ' Why – only one. Why should they take more? They will leave here in the morning while it is yet starlight: during the day we will

Below. *Striking camp on a clear crisp morning on the Nyamugasani route in southern Rwenzori. In the background is Mount Rwatamagufa (13,953 feet) – 'He who breaketh your bones.'*

move to Nyabitaba, and they will meet us there tomorrow evening.'

I was almost incredulous, but I could not doubt them. Even when the Konjo are heavily laden, their capability to travel is amazing enough: the daily marches fixed by Europeans give a quite false impression of their true ability. But unladen they travel like the wind.

So it was decided. Muthundya came to my tent before sunrise and my letter, wrapped up in a dry groundsel leaf to look like a twist of tobacco, was tied into a knot in the corner of his blanket. I wished both him and Bukombi God-speed and silently they were gone. We broke camp as the first sun flooded the valley and then, even as we grasped our staves to set off, the mountains reverberated for several minutes with the chilling sound of gunfire from the direction of Ibanda: not the massive explosions of yesterday, but a continuing drumfire. I guessed it was a mixture of 40 mm cannon and small-bore machine guns. No one said a word: we just looked at each other, shrugged our shoulders and set off.

We were back at the familiar campsite on the ridge by midday and I spent the afternoon trying to anticipate what I might be in for. At the least, I must expect to be interrogated and my kit searched and pilfered. I had with me a number of papers on the Ugandan and Rwenzururu situations, all of which could be regarded as seditious by an ill-disposed person. I also had, in paperback, a copy of Professor Kenneth Ingham's book, *The Kingdom of Toro* – clear evidence that I was a dabbler in politics! Worse, I had my set of maps of the range: maps are universally associated with espionage and malfeasance in Africa. I wrapped all this incriminating matter in polythene and buried it at the foot of a podo tree. My main concern, however, was for my exposed film cassettes. I transposed the labels between used and unused, and hiding the used in my rice, packed the unexposed but apparently used films where they would readily be found.

Darkness crept over our ridge, and I was in a

state of mounting despair when, out of the night, I heard a gentle voice.

'*Nipo mimi, bwana. Tumerudi* – it is me, sir. We're back.'

It was our young Phillipides, Muthundya, his marathon run, and behind him his shadow, Bukombi. I almost wept with relief. Fumbling in his blanket, Muthundya produced a crumpled letter from Moses, written on the reverse of mine.

'All is safe and no one has been harmed. You will please be welcome to come here tomorrow and rest after your troubles.'

I shook Muthundya and Bukombi warmly by the hand.

'You have travelled three days' journey in one,' I said, 'and so have earned three days' wages: but my heart is melted by the good news you have brought, so you will each receive six days' pay.'

They both had enough energy left for a jig of delight: I believe they had no idea of what sort of feat they had performed by European – or Greek – standards. I sent them off to the rock shelter with the letter for Erinesti to read to the others, and then rather shamefacedly disinterred Professor Ingham! In my sleeping bag, gloriously relaxed, I enjoyed a chapter of Trollope (I was reading *Framley Parsonage*) before settling down to sleep. From the Konjo camp came the sound of laughter and singing: all was well at last.

At Ibanda they were as puzzled about the explosions and gunfire as we were, but it had come from further south, and was presumed to be an attempt to terrorise the irredentists. That night three or four leading villagers asked if they could discuss the situation with me and we met clandestinely in one of their houses by oil light. The cause of this resurgence of government activity, they told me, was the problem of the coffee crop. Successive governments had tried to control the sale of the crop, and so maximise revenue, by fixing the price to the producer and enforcing marketing through co-operatives. Not only was the price too low,

but payment was unreliable. Meanwhile their brothers in Zaire enjoyed the blessing of a well-managed free market. Naturally the Ugandan crop found its way to Zaire, humped across the mountains on Konjo backs. For themselves, they will carry twice the weight that they will carry for me – easily – 100 pounds! The best route is across the Tako river, through Rwenzururu territory. The reaction of the rebels was to demand a duty on passage. This in no way restricted the trade, while the kingdom was revivified by a surge of revenue.

'How are they paid?' I asked.

'In Uganda shillings, of course – nothing else would be of any use to us.'

'How is it that the Zaireans have such currency? It is illegal to export it.'

They looked awkwardly at each other and then spoke in lowered voices.

'You know what Kampala's like. It's full of Big Men – they have power and big cars and travel where they like. And then there's all these dollars that come in as aid. Where do you think all that goes? But it comes in through the banks and these men can only get hold of it as Uganda shillings – and these are worthless in Europe. But in Zaire they can buy gold with it. There is any amount of black market gold in Zaire. So they get this gold and when they go on business trips to Europe their briefcases are full of the stuff and it all ends up in Swiss bank accounts. So Zaire has this supply of Uganda shillings and they can buy our coffee at a fair price.'

One of the villagers mentioned to me that there was an ancient smugglers' route that started in another valley and took one right over the range to Zaire. I was still smarting over my rebuff in the Kuruguta. I had supplies and eager porters: a new (to me) route via an old smugglers' path suddenly seemed attractive and on the spur of the moment we decided on the trip. It was an excellent decision. Two days later, with twelve porters under an old (and avowedly retired) smuggler, Muhindo, we traversed the foothills through delightful forest and made our way up a bracken-covered ridge to a lovely campsite at 8,000 feet. Our next two camps were at fine caves in a narrow valley and on the fourth day, having sent the porters down to a lower rendezvous, Muhindo and I climbed steeply into the dramatic cleft of an unnamed pass at over 13,000 feet. A beautiful lake lay below, for all the world like a landscaped lake in English parkland, and beyond, the measureless forests spilling down into Zaire.

It only took us a few days, downhill, to return in fine weather and as we paced easily along I discussed this unexpected route with my smuggler companion. I had noticed that some of it had been better aligned than one generally expects in Rwenzori, and parts even had the appearance of having stones set to give an improved surface.

'It is a route of great antiquity,' said Muhindo, 'our grandfathers told us about it, and theirs before them. In those days we were all one people, the Banande and ourselves, and naturally we had a route right across the mountains, for trade, and visiting and seeking wives, long before the Europeans divided Zaire from Uganda.'

Once the motor road round the south of the range to Zaire was made – the so-called Equator Road – the high mountain routes fell out of use except for hunters. But the closure of the frontier and the rise of coffee smuggling had brought them back into heavy use. I had noticed that the floors of the caves were covered with spilt coffee beans, which made them comfy to lie on.

Down in the forest we came to a rather nice level area where Muhindo showed me several immensely trunked trees, one or two of which had entrances to hollow interiors big enough to provide sleeping space for two or three men. These floors too were covered in coffee beans.

'In the old days,' Muhindo told me, 'before the British Forestry Department evicted everyone, there were always a few people who had their homesteads high in the forest. Since time immemorial they had planted these *orubugu*

trees, from which they made bark cloth.'

The trees are of the wild fig family and are certainly hundreds of years old, so this was a marvellous window looking back on to the history of Uganda. Before Arab traders started bringing cotton goods into the country, the people dressed in bark cloth. This is a sheet-like web of brown fibres that can be peeled from the trunks of the trees. If beaten with wooden mallets, its area is extended, as gold leaf is made, and the resulting material is flexible and can be handled like ordinary cloth.

'Now of course nobody lives here,' Muhindo went on, 'but the trees make good shelters for smugglers. There are no rock shelters at this level, and as they usually only travel in ones and twos they can sleep dry inside with their loads.'

Fire ashes confirmed what he said, but I remarked to him that on our present journey we had not met any smugglers.

'Everyone got wind of the new NRA offensive,' he said, 'so they are all lying low until it has blown over. But you can be sure the path will come back into use – it's one of the ways by which the Big Men get their gold from Zaire!'

On the last day of this little side trip, my heart sank as we met a patrol of the Resistance Army coming up the path. There was a corporal and half a dozen men dressed in dirty camouflaged denims, all carrying the venomous Kalashnikov automatic rifles that have done so much to destroy Africa. They stared at my wild and bearded form but did not speak to me and I passed on my way with despair in my heart. In due course the army established a permanent post at our lovely campsite at Nyakalengija, and the innocence of that little spot was lost.

VOLCANOES OF THE MOON

Which are the Mountains of the Moon? Stanley's Rwenzori, Speke's Mfumbiro (Virunga), or Baumann and Kandt's Missosi ya Mwezi? In 1956 I had stood on a green crest at 7,000 feet in southern Burundi with a Belgian veterinary colleague, and he had dramatically extended his right hand to the south where Lake Tanganyika lay in its immense trench and exclaimed, 'Voilà – c'est le Congo!' and then, turning to the north and extending his left hand, 'Voilà – le Nil d'Egypte!' In the early months of 1987 I returned to that same spot in Burundi and thence, slowly working my way up what they call the Crête du Nil – the Nile watershed – found myself in Kigali, the tiny capital 'village' of the little country of Rwanda. My objective now was to complete my observations of the afro-alpine regions by visiting the cluster of 'sky-scraping cones' that Speke, in 1863, had proposed as the Mountains of the Moon.

I engaged an African driver and we drove west through the astonishing complexities of Rwanda's entrancing mountainous countryside – the *mille collines*, as they call it – until, on rounding a bend on a steep hillside, I saw, misty purple in the distant haze, a moonscape of cones like the background of a Japanese watercolour. As the sun set behind them, we made our way to Kinigi, which is the headquarters of the Parc National des Volcans. That we had entered the volcanic region was now all too obvious, for the road was abrasively rough with laval debris, while a fine grey volcanic powder permeated the car. We were gaining altitude, gently ascending the outer volcanic fan-slopes to the park headquarters, where there is a small office, some bunk houses and a grass enclosure where I pitched my tent. This is an open treeless area, until recently virgin forest, once sacrosanct but now excised from the park.

Those who believe that national park status necessarily provides inviolable protection should study the sad history of Rwanda's Parc National des Volcans: between 1958 and 1979,

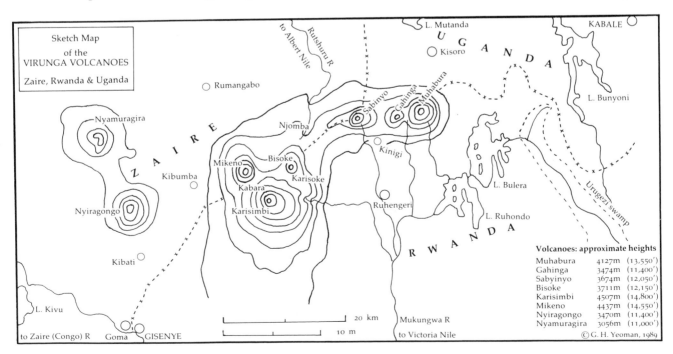

Sketch Map
of the
VIRUNGA VOLCANOES
Zaire, Rwanda & Uganda

Volcanoes: approximate heights

Muhabura	4127m	(13,550')
Gahinga	3474m	(11,400')
Sabyinyo	3674m	(12,050')
Bisoke	3711m	(12,150')
Karisimbi	4507m	(14,800')
Mikeno	4437m	(14,550')
Nyiragongo	3470m	(11,400')
Nyamuragira	3056m	(11,000')

© G. H. Yeoman, 1989

over half the original park area was surrendered for agriculture. This area in fact included almost all the lower broad-leaved forests. It was a response to the ever-rising pressure of the human population, and the particular aim was the production of pyrethrum (an insecticide-yielding daisy). The funding of this massive destruction of one of Africa's most beautiful and sensitive ecosystems came from our own European Economic Community – that is, from my pocket, and very likely yours. It was a misconceived development project that not only destroyed the environment but was destined for economic failure because of the entirely foreseeable development of semi-synthetic pyrethrum substitutes. The environmental destruction was especially tragic because by thus drastically reducing the rangelands of the small surviving populations of the mountain gorillas, it probably sealed their fate. It was my feeling that the history of the Virungas, in the face of population pressure, could provide a forewarning of what could happen to Rwenzori, that had brought me here.

There are five great volcanoes on the Rwandan border and a further three inside Zaire. They stand in three groups: from east to west the first group are named Muhabura (13,550

Below. The truncated cone, right of centre, is Mount Bisoke (12,150 feet), main stronghold of the mountain gorilla in Rwanda. Distantly to its right, emerging through the cloud, is the summit of Mount Mikeno (14,550 feet) in Zaire, and to its left, cloud-capped, is Mount Karisimbi (14,800 feet). In the foreground are fields of daisy-like pyrethrum, part of an ill-conceived EEC-funded development scheme that required the destruction of much of the national park's forest, and which has proved an economic failure.

feet), Gahinga (11,400 feet) and Sabyinyo (12,050 feet), and these stand on the border with Uganda. But Sabyinyo is tripartite, since on the highest of its several summits the borders of Uganda, Zaire and Rwanda meet. Further west is the handsome second cluster of Bisoke (12,150 feet), Karisimbi (at 14,800 feet, the highest), and Mikeno (14,550 feet). The first two of these stand astride the border between Rwanda and Zaire and are classic cratered cones, but Mikeno is entirely within Zaire and its spectacular peak is a caldera remnant. None of these six volcanoes retains any activity, but the remaining two, which are both within Zaire, are most decidedly active. These are Nyiragonga (11,400 feet) and Nyamulagira, which at 11,000 feet offers the lowest profile of all. Graf von Götzen, when he first explored this region in 1894, found Nyiragonga dramatically venting and lighting the night sky, and so it has remained, on and off, ever since. Both these volcanoes are continuously smoking and smouldering and each has produced a disastrously damaging eruption in recent years – Nyamulagira in 1938 and Nyiragonga in 1977.

These then are the Virunga volcanoes – the Mfumbiro of the early explorers and treaty-makers. They lie more or less in line astride the Rift valley and effectively block it, creating three watersheds. Two of these are to the Nile: east and south-east to Lakes Bulera and Ruhondo and so to the Kagera Nile and Lake Victoria; north to the Rutshuru river, the Semliki and Lake Albert. Otherwise, the drainage is west and south to Lake Kivu, Lake Tanganyika and the Zaire (Congo) system.

I had come with the intention of climbing as many of them as might prove possible, but now that I was here and looking up at the uncompromising cones, my heart had some misgivings. However, there is nothing to be done about unpromising enterprises but make a start, so I engaged a park askari as a guide and a couple of boys as porters. We would begin by tackling the two eastern peaks, Muhabura and Gahinga.

As I had insisted, the men came to my tent before sunrise – three sanguine-looking Banyarwanda, with frank, chocolate-coloured faces and short, sturdy limbs. John, who was an accredited national park guide, had a middle-aged, weather-worn, and a reliable look about him. The other two, James and Gerard, were innocent teenagers who had begged me the previous evening to take them as porters. Bearing in mind that we were proposing to climb two high volcanoes and sleep two nights on the mountains, their equipage could be described as minimal. Each was bare-footed and dressed in cast-off trousers and shirts in an advanced state of disintegration. The only sophistication was provided by John, who had a shabby ex-army combat jacket and an old canvas sleeping bag which he proudly announced as being 'big enough for two'. He also carried a time-worn and questionably serviceable .303 Lee-Enfield rifle, with three rounds of service ammunition in his pocket – 'in case of buffaloes or things', he said darkly.

We had about 4 miles of gently rising volcanic slopes to cross before we could even turn and face our mountains and start the serious ascent. After an hour we came to a village of tumbledown mud and lava buildings. Here my men announced that they had to buy rations for the trip and disappeared into what was obviously a beer house. When they emerged half an hour later, carrying what seemed to me to be an inadequate supply of small round loaves wrapped up in a grass bundle, they were wiping their lips and offered me rancid-smelling liquor, to be sucked from a Johnny Walker whisky bottle through a communal hollow stick. In happier times I would have accepted and drunk deep enough to seal our camaraderie: but now, hating myself, I politely declined. Was this not the Africa of pandemic AIDS? To cover my embarrassment I sent them back to buy a few pounds of rice: they should at least have full bellies while in my service.

Bright sunshine had broken through the mist

as we made our way up amongst the highest gardens towards the edge of the bamboo forest. But distant thunder from the still-invisible peaks presaged black clouds overhead. A scattering of heavy premonitory raindrops warned me to put on my anorak, but to no avail, for within a few minutes the deluge hit us, a combination of rain and hail of devastating ferocity that had us soaked to the skin in no time, my boots overflowing with water. This had the effect of bringing out every kind of jollity in my companions: high-pitched exclamations, bursts of laughter, scraps of song – what to me was infinitely depressing seemed to them to be the most amusing thing. But not for long: after ten minutes or so there was an ear-splitting explosion which caused James and Gerard to topple my loads in the mud: no flash, but evidently a near-miss lightning strike that silenced them for a little while before they regained their equanimity and retrieved the loads, and we proceeded on our way.

Our path now became a buffalo-tramped sloppy morass of volcanic mud, and even when the rain at last slackened, the bamboo fronds continued to shower us with water at the slightest touch. The gradient was tedious but by five o'clock we emerged into more open, grassy vegetation, with picturesque hagenia trees and golden-flowered St John's wort, and my companions reverted to their high spirits. We were climbing on to the broad saddle, about 9,000 feet in altitude, which connects the queen of the volcanoes, Muhabura, with her lesser sister Gahinga. The Rwandan-Ugandan border runs through their two summits, and almost on the putative international boundary line, we came across the dilapidated metal rondavel – a small round hut – that was to be our sleeping quarters for the next two nights.

Cold and wet as we were, a fire was our top priority and tea the next. My Australian waterproof matches saved the day, and while John got a fire of dry bamboo kindling going, I went with Gerard in search of water. Just below our camp was a spring of Guinness-coloured water,

emerging from an outcrop of volcanic tuff into a little pool. The spring was unusual: springs and streams are rare on these mountains, as most of the water is absorbed by the volcanic material, to emerge lower down where it meets an impermeable stratum. It was a special delight to me – a true high source of the Nile: being just south of the divide, its waters were destined for the enchanting Lake Ruhondo that we could see below us in the light of the setting sun, thence across Rwanda to the Kagera and so into the Victoria Nile. The thought pleased me, and I sipped my tea with extra pleasure.

I stirred my bedfellows early and emerged from the hut, rubbing the smoke from my eyes, into a world of dense mist that was only just beginning to lighten. As I looked towards the Ugandan side of the saddle, strange dark shapes began to appear. At first I thought of buffalo, but these were rounded lumpish shapes, which soon declared themselves as bulging sacks, each walking on two legs, the rest of their human carriers being invisible. I quietly called John and he made haste to join me with his rifle, followed by the other two. The sack-carriers – there were six of them, bowed down and preoccupied – were quite unaware of us until – ambush! – John pounced. With dismay they dropped their loads, for a

Opposite left. Helichrysum stuhlmannii, *the everlasting* immortelle, *is found over a wide altitude range from 10,000 feet to above the snow line, but it is in the higher belt of this zone that it dominates the vegetation. Its olive green and silvery foliage is adapted to control water loss in the critical hour immediately after sunrise; its silver flowers are nearly always closed in cloudy weather, but open wide as soon as the sun appears.*

Helichrysum guilelmii *is a less hardy plant, and its more fleshy lax leaves rely for protection against the cold and desiccation on a system of closely woven 'spider's web' fibrils. The delectable rose-tinted flowers are found on the higher altitude specimens, the lower altitude varieties tending towards a creamy silver.*

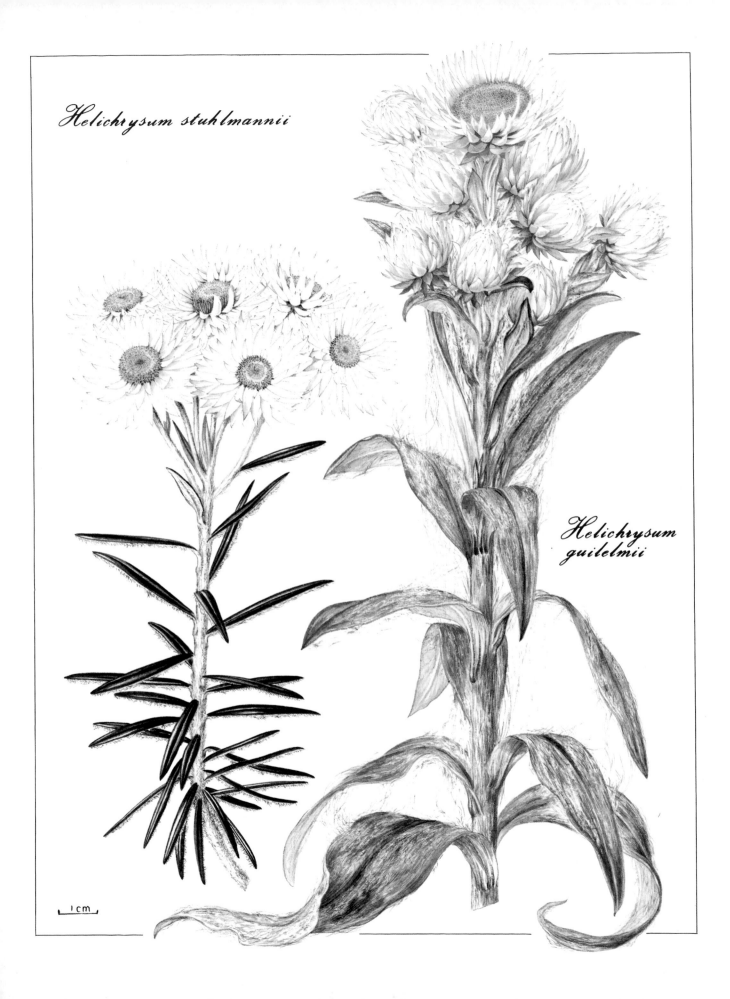

Helichrysum stuhlmannii

Helichrysum guilelmii

1 cm

moment contemplated flight, and then fell silently to their knees.

John looked at me delightedly: 'Tchaa – ah! – smugglers,' he said, 'and we've caught them red-handed.' It certainly was a fair cop, and while he at rifle point (though I noticed he had not bothered to load) made each man produce a smudgy, rain-soaked *carte d'identité* from amongst his scanty rags, I inspected the much-patched gunny bags. Each was crammed with bright red millet, and weighed not less than a hundredweight. My main concern was of delay to our start for Muhabura, but I was also anxious to take any heat out of the incident: when the officious John asked me for pencil and paper to write down the miscreants'

particulars, I made it an excuse for us all to go back into the hut where at once, their natural good manners reasserting themselves, my men made up the fire, settled the shivering smugglers around it and passed round a cigarette stub, while John put me in the picture.

'It's like this: in Uganda there are Banyarwanda who grow millet, but such is the mismanagement in that country that they are forced to sell it for almost nothing. Now here in Rwanda it's too cold for millet, but of course, as you know, millet is needed for brewing the best beer, so naturally there is a trade across the mountains. It's illegal, because they ought to pay tax. So these people are smugglers and must be dealt with severely. The question is, how

Below. *Viewed from near the summit of Mount Muhabura (13,550 feet), the cratered summit of Mount Gahinga (11,400 feet). Behind it rises Mount Sabinyo (12,050 feet) which I climbed with my four women companions and Mzee Zacharia Ngango in 1987.*

best to do this?' He looked at me hopefully.

'Perhaps we had better make tea,' I said, rummaging in my pack for the makings, while Gerard was sent for water and James added kindling to the fire. 'Then we can think about it properly.'

An animated argument now ensued between John and the chief smuggler in the Kinyarwanda language, incomprehensible to me, but increasingly looks came in my direction. I noticed that John's hectoring slowly moderated, while the cringing smugglers, humble men with wrinkled but far from wicked faces, began to assert a little confidence. At last, as my single tea mug was passed round, John announced, 'Well, M'sieur Guy, we know you are an Englishman and understand matters of justice, so we are all agreed to ask you to decide the case.'

Me! Twenty-five years after independence, in a far country where the writ of British magistracy had never run, the Englishman, whether in the guise of Solomon or Pontius Pilate, was turned to for wisdom and justice! I knew that behind this little rustic drama lay one of the deepest ills of Africa. These new states, structurally enslaved to inappropriate Western-style fiscal philosophies, with their ministries, banks and international loans, their city *nouveaux riches* and corruption, have universally strangled their only true and sustainable source of wealth – their peasant farmers – by artificially depressing the price they get for staple crops in order to subsidise unproductive urban life and governmental squandering. Who is the law-breaker? The ruler who denies the labourer his just rewards, or the labourer who takes what resort he can to feed his family?

I called the meeting to order and prepared to deliver my summing up, my first problem being that of language. English would be as incomprehensible to them as their Kinyarwanda was to me, but both Swahili and French are widely spoken in the region and it was in a French-salted Swahili patois that I spoke. The way in which ideas are expressed in Swahili by rural Africans always seems to me to have a Biblical simplicity about it, and in reproducing conversations I have tried to suggest this poetic quality of the language. I fixed the smugglers with a stern eye and, emphasising my words with an extended right hand, I spoke out.

'Look here – now, listen all of you. You've been caught in the act. Are people to be allowed to break the law like this? If the government is to be robbed of its taxes, why there would be no end to it and we should live in chaos. John here is quite right and only doing his duty as an askari. The fact is, you should all be marched down to the police post at Ruhengeri, thrown into gaol and your millet confiscated.'

The smugglers hung their heads wretchedly and John smirked self-righteously, while James and Gerard looked at me with open-mouthed admiration. I paused, and then, extending my left hand, went on in a moderated voice. 'But on the other hand, it cannot be denied that poor men work hard for a living and get but small reward. People must have millet for brewing beer – what would our lives be like without it? – and it may be argued that those who sweat to carry it over these great mountains in order to maintain a cheap supply are doing us all a favour for which we should be grateful. After all, if only the Big Men at the top were more sensible, such labours would be unnecessary.'

The effect of these words on our captives was as water on wilting plants: they straightened up and filled out, and with sparkling eyes, cried out, 'Every word he says is true . . . a true *gentilhomme anglais*,' while even John, his *amour propre* sufficiently placated by my preamble, joined in the approbation. I ostentatiously looked at the doorway, through which light was now shining.

'I propose that each of us should now set about doing whatever he has come here for: for me and my companions, it is the ascent of Muhabura – that is all that concerns us. As for these others, they should take up their loads and depart, and may God bless their journey.'

Each man pressed my hands, and John,

James and I, leaving Gerard to keep camp, set off up the mountain with strangely light hearts.

Climbing Muhabura was misery. Desperately steep, it presented the anomaly of a near-vertical mud wallow compounded by intractably tangled, soaking vegetation. My climbing boots were positively dangerous, their rugged soles sealing over with mud, and the wisdom of my companions in going barefoot was all too obvious. The vegetation of lobelias, groundsels and hypericums showed the familiar giantism, but it was less distinguished than on Rwenzori, only the groundsels being impressive with their heavily flowering, golden spikes.

Below. *The beautiful flowers of the Rwenzori blackberry*, Rubus runssorensis.

Not far below the summit, a vertical cliff forces the climber to traverse clockwise, and to his relief he finds himself on reasonably good rock until from the Ugandan side a pleasing grassy ridge leads to the summit. This must be a lovely place in clear weather, since the open crown of the mountain has a perfectly circular miniature crater lake of pure water, perhaps 50 feet across, artistically landscaped with a few giant lobelias. But for us the freezing Scotch mist was so dense that we could barely see across the lake. It was cold, grey and gloomy, and after sharing a few groundnuts and raisins, we hastened to make our descent. In Zermatt the Swiss say, '*Mont Cervin, il fume sa pipe*', pointing to the apparently innocuous whisp of cloud on the Matterhorn that in fact can spell such misery for the climber. So it is with the Virungas. For the complacent voyeur on the plains below, their cloud caps give artistry to the scene: indeed, when rarely they are cloud-free, they lack character. But when you are up there, such clouds can offer a potent vortex of hail and bitter, storm-force winds, and of course they rob you of any view over Africa.

By contrast the ascent of neighbouring Gahinga next day was delightful. It is a shorter climb than Muhabura, and most of the route has a fairly good base of laval tuff, although this can be slippery when wet. The buffaloes seemed to prefer this mountain, and had cleared a path for us through the mimulopsis and nettles, while the bright sunshine was tempered by the heath and hagenia trees.

These Virungas are like the pyramid mountains that children draw, and the transition from the comparative flatness of the saddle to the uniform steepness of the slope is sudden. Within a few paces one is grappling with a despairingly steep tangle and progress is largely made by hauling on vegetation. You soon learn what you can trust: there is nothing sturdier than the ericas, root or branch; senecios and hypericums can usually be relied on; but oh! – never put your trust in a lobelia. Such ascents have little appeal to the simple mountaineer, but to the naturalist and to those seeking to explore their environment, they have much to offer. I have never tired of the uplifting transition from the hot plains of Africa to the ethereal heights. With every step upwards, be it never so weary, something changes, as the familiar plants give way to the afro-alpine world, and with every breath the air becomes more delectable.

Gahinga is particularly distinguished by its blackberries. These are a feature of both Virunga and Rwenzori. In their favoured patches they grow exuberantly, with sweet tudor-rose flowers and fruit in all stages of ripeness on the same stem – presumably a reflection of the absence of seasonality. And what fruit! Two or three times the size of their European congeners, they have a bitter-sweet juiciness that is totally thirst-quenching, the final answer to the inescapable dry throat of high climbing. John and Gerard were ideal companions, easily settling for my measured pace, while supplying me with a life-support line of blackberries.

After only two or three hours we took a pace or two on to open tussock grassland, and there was the crater at our feet – in fact, two craters, one terraced above the other, both brilliant golden-green in colour, each about 200 yards across. They are dangerous, bottomless bogs, and their colour comes from the vivid sphagnum mosses and sedges with which they are covered. Surrounding them the rim of the

Below. *On our descent from Mount Sabinyo, the mist lifted to give us this view of Mount Gahinga, framed between a rapanea tree on the left and a hypericum on the right.*

caldera is a delightful sward picturesquely set about with silver-flowered helichrysum (the Rwandans' *immortelles*) and giant-branched senecios and lobelias. Especially striking were masses of Stairs' deep-rose-coloured ground orchids, that were familiar from Rwenzori.

On this day at least, we had our view. In Uganda we could see Lakes Bunyoni and Mutanda, nestling like Italian lakes in the wrinkled hill country of Kigezi, indicating the Lake Edward and Semliki Nile catchment, while Rwanda's maze of cultivated *mille collines* and swamp valleys extended beyond vision in the haze to the south, to the Kagera and Victoria systems. Here we stood precisely on the divide between Lake Albert and Lake Victoria. To the east we had a clear face-on view of mighty Muhabura, scene of our toils of yesterday; to the west, many-toothed Sabinyo, with which I yet had to come to grips. Down on the saddle below us, a pinhead of reflected light showed us our hut, with a thin smudge of blue woodsmoke.

'That idle fellow James,' said Gerard, 'he's just dozing there by the fire, never thinking of fetching wood or water.'

Overcome by high spirits, he climbed up into the branches of a giant groundsel and asked me to take his picture. As I complied, I said to John,

'This is excellent – when I show it to my friends in England, I shall tell them it is the gorilla of Gahinga.'

Gerard was convulsed by my wit, and as we tumbled joyfully down the mountainside, kept repeating my words.

'The gorilla of Gahinga – jee! Me! – a gorilla!'

One is tempted to say that when you have climbed one of these volcanoes you have climbed the lot: there is some truth in this, but it is not quite fair. Over the next few weeks I climbed most of them – indeed eventually, all but Nyamulagira – and each was a sufficiently different experience to make the enormous physical effort worth while. Bisoke is probably the most frequently climbed, because of its accessibility. It is a steep and demanding haul, rewarded, if cloud allows, with a view into a pretty crater lake.

Bisoke's neighbour, Mount Karisimbi, Rwanda's highest peak, is considerably more demanding. I took two days over the climb. You must first make your way up through a narrow rock gulch and a densely vegetated valley of bamboo, hagenia and hypericum forest. After a couple of hours this brings you on to a more or less level saddle at about 10,000 feet between Bisoke and Karisimbi, at the place where the American gorilla protagonist, Dian Fossey, established her study station in 1967 and met her violent death in 1985. She named it Karisoke, after the two mountains. When the sun is shining, which I suspect is not very often, the saddle is a pleasant place. But perhaps because I knew the history of its recent tragedy, the green-painted corrugated-iron huts created a melancholy impression; the more so because they were temporarily unoccupied; and the graveyard, gorilla and human, which lies nearby under dripping trees, does nothing to dispel the gloom.

My African companions and I did not loiter, but turned towards Karisimbi and pressed upwards through the dense and soaking tangle. By about midday, we reached steep but more open ground in which we found a rather wretched triangular pig-ark of a hut, where we dumped our loads and made tea. By now the cloud had lifted, the sun was warm and the summit – only two or three thousand feet above – tempted us to go for it straight away, rather than risking a bad day on the morrow. So warm was the sun and so fair the conditions, that I foolishly just took my anorak and cameras, and nothing else.

All went well for the next hour or so, and we enjoyed splendid views over the whole volcanic chain. But then the cloud closed in, the wind got up, the temperature dropped alarmingly and I found myself caught out, insufficiently protected, in one of the violent vortex storms that are characteristic of these peaks. The cyclonic freezing wind was indescribable and time and again we were simply thrown bodily to the ground. First we endured horizontal icy rain, then painfully cutting hail, and finally a penetrating granular snow. The surface of the mountain, featureless at the best of times, became a vast snow-drift which with the dense cloud created a perplexing white-out disorientation. Warmth and energy were sucked from my body; hands and feet froze; my eyes became glazed, unfocusing, and my mind numb.

My companions, less well clothed than me, suffered all this with the utmost fortitude, and it was only shame that kept me from turning tail and making haste down to warmer altitudes. At last we reached the summit – a hideous slag heap of a place, temporarily improved, I have no doubt, by the snow, but even so sadly cluttered with the junk of a derelict radio mast which clearly had not been constructed with Karisimbi's hurricane-force winds in mind. Rwanda's highest summit is a disgrace and should be thoroughly tidied up. But I was too miserably cold to take much of it in, and we soon set off downhill, to the hut and life-saving tea.

Only later did I read about the tragic experience of the German geologist Kirschstein on Karisimbi in 1907. After making an ascent for scientific study with a large party of inexperienced porters, he was caught out in a violent vortex. Africans, who can be so amazingly stoic in circumstances that are unendurable to Europeans, can equally, when fate seems to be immutably against them, abandon hope and literally give up the ghost. So it was with Kirschstein's party, and twenty of them perished, minutes rather than hours away from safety. Kirschstein himself developed pneumonia and lay unconscious for two days. Such can be the unpredictable violence of these realms of snow on the Equator. How thrilling it is to think of the placid Nile of Egypt being wrought into existence with these traumatic birth pangs.

The entire chain of volcanoes is about 50 miles long and although it is practicable to climb the five Rwandan cones from the national park, I missed Sabinyo out on this occasion and only got round to climbing it some months later when, with my English women companions, I was searching out the headwaters of the Rutshuru river and so of the Semliki Nile. From the last little township of Uganda – Kisoro, a down-at-heel dusty place in an impoverished lava landscape – we followed a really terrible track down into the south-westernmost corner of Uganda, ignoring the tearful protests of our driver. It took us two nights of camping before, finally running out of track, we felt ourselves to be within striking distance on foot of Sabinyo.

On our way we had picked up the *Mzee* Zacharia Ngango, a time-weathered, wise and responsible game-guard, at his picturesque thatch house at Rukungi at the foot of Muhabura. As far as I could see, Old Zacharia represents the sole remaining evidence of Uganda's commitment to what was once the Ugandan Gorilla Sanctuary, which was continuous with those of Rwanda and Zaire. Now all the forest has disappeared right up to the bamboo line on the very cones themselves – an awful warning for Rwenzori – and it is questionable whether any truly resident population of gorillas remains.

With Zacharia and two young porters, we started from camp at first light, trudging for 2 or 3 miles over unpromising laval scrublands which were surprisingly bedecked with freely flowering sorrel, before entering the bamboo and mimulopsis thicket and starting up a lava ridge which soon increased in steepness, developing deep chasms on either side. Progress was slow and exasperating – wretchedly slippery soil, rocks, roots and tussock under dripping hagenia trees. Higher up there is a curiously uniform and unattractive zone of thin wind-pruned giant heath, evenly bearded with dismal grey Spanish moss. Above this, the ridge became even steeper and narrower, and alchemillas, helichrysums and lycopodiums made their appearance. Often we were reduced to our knees, hauling ourselves up by clutching the unreliable vegetation. But all this was nothing to the women, who had been case-hardened by a month on the Rwenzori, and they managed it every bit as well as the Africans.

Sabinyo has several distinct and sharply pointed summits – visitors' counts seem to vary between five and seven. They are the drastically eroded remnants of a once-deep caldera, and this mountain provides a more dramatic and complex landscape than the other volcanoes. Having at last reached the top of one of these summits, Zacharia informed us that it was only number three, and we had to traverse a second before reaching the true summit. On our way, we at least had the satisfaction of noting clear evidence that these knife-edge cols between the peaks had recently been crossed by gorillas. A group of gorillas produces a characteristic swathing of the vegetation as it goes on its way, and leaves further evidence in the form of copious quantities of tri-lobed, horse-like dung.

On the true summit, although we were cloud-veiled, we did have some striking views down on to the still largely undamaged forests of Zaire, and on to the familiar rectangular pyrethrum fields of Rwanda. Sharing our rations, towards which Zacharia's wife had thoughtfully contributed cold boiled beans and potatoes, we took some satisfaction in debating in which of the three countries that meet at this point, each of us was lunching. The clouds closed in as we descended, but the rain held off until we emerged from the bamboo forest at dusk, when the skies opened, drenching us long before we reached camp in the dark.

My purpose in visiting the Virunga had been to learn what happens when the pressure of a growing population impinges on mountain forests in circumstances where there is inadequate national will to protect them. I did this deliberately in order to give my voice authority in warning of the fate in store for Rwenzori. Even so, I had not been prepared for the shocking situation I should find. Both Rwanda and Uganda have given their beautiful volcanoes an uncompromising 'short back and sides' that has immeasurably devalued them. What had the Zaireans done on the other side? I made my way across the border, determined to find out for myself.

From the frontier town of Goma, on the north shore of the entrancing Lake Kivu, which I think is the most beautiful of all Africa's Rift valley lakes, Nyiragonga's cone dominates the scene, 12 miles to the north. Since Count von Götzen's day, most visitors have written of it as painting the night sky red with its flaming crater, while by day its column of smoke guided them from afar. This has disappointingly ceased to be the case since the last devastating eruption in 1977, although when you are actually on the mountain there is still a good deal of sulphurous activity to be seen and smelt. I camped at its foot and next day set out with two local men on an ascent that proved to be a wearisome and gloomy experience. The slopes are made of laval rag, with an impoverished pioneering flora of which ferns surprisingly made up a high proportion. After four hours we came to a slightly less steep scallop in the mountainside where there were three vandalised aluminium rondavels. This mountain used to be a popular ascent when there was a thriving European population in Goma (long since departed), especially when there were night-time fireworks to be seen. Two of these huts are now unusable, but we made ourselves as comfortable as we could in the uninviting third, and got a fire going. At two o'clock in the afternoon we were already in dense, cold cloud. My companions roasted maize cobs and boiled potatoes, kindly providing me with a hot lunch.

By four o'clock the cloud showed signs of breaking up, and my guide and I set off straight up the final few hundred feet of nearly bare volcanic debris to the crater's rim. The caldera presents a truly fearsome sight. I found it hard to estimate the dimensions, but I guessed it was about half a mile in diameter, perfectly circular with an almost uniform rim, and between 1,000 and 2,000 feet deep, with vertical walls. At the bottom was a lake of congealed, grey-black, metallic-looking lava of repulsive appearance, from which sinister fumaroles vented white sulphurous fumes.

Gloomy at any time, on this grey, cold and cloud-swept evening, with the invisible sun setting beyond the Zairean Rift valley wall to the west, the prospect was particularly forbid-

Below. Mzee Zacharia Ngango, sole remaining guardian of the now deforested Ugandan Gorilla Sanctuary on the northern flanks of the Virunga volcanoes. He was our guide for the ascent of Mount Sabinyo. The portrait reveals his gentle, wise and trustworthy character.

ding. We walked a little way round the caldera edge anti-clockwise, but as we turned back, somewhere to the south the cloud must have opened, for the sun broke through and we saw Lake Kivu as an intricate fairy-tale sheet of silver, suspended in the mist. Grateful for this gift, we stumbled down to our hut, and made the best we could of our night on the bare mountain.

From the summit of Nyiragonga, looking north-west, we had had thrilling glimpses of Mount Mikeno. This was something I could scarcely ignore, a veritable tropical Matterhorn. As the evening cloud lifted, I could see that the mountain and its brother Karisimbi, both snow-dusted, rose above a dark and extensive apron of forest which spilled out into the Rift. So it was true: the Zaireans, all credit to them, had so far maintained the integrity of their famous Parc National des Virunga. I knew that these dark forests provided the only real hope for the survival of the mountain gorillas. So special are these animals and so unusually dramatic the story of man's attempts on the one hand to destroy them and on the other, to save them, that I felt I could not leave this region without making an attempt to visit them.

Overleaf. Lac de la Lune, 13,300 feet, is one of the most northerly of Rwenzori's lakes and possibly the deepest. Lying tightly between the two northern-most snow massifs, Emin and Gessi, it seemed to Dr Noel Humphrey's to meet the description of Herodotus (450 BC), that the ultimate source of the Nile was a bottomless lake lying between two peaks, Crophi and Mophi. It is the source of the Ruanoli stream that flows into the Semliki river and so into Lake Albert.

GUARDIANS OF THE NILE SOURCES

My view of Mount Mikeno's 14,550-foot peak from the rim of the smoking crater of Nyiragonga had determined me to attempt an ascent, although I knew that without a rock-climbing companion the summit would be beyond my reach. A few days later I arrived at Kibumba at the foot of the mountain and engaged three young porters and two elderly guides – a Tweedledum and Tweedledee partnership – I could either take both or neither! Semajeri and Kokota were even more aged and wrinkled than Zacharia had been on Sabinyo, but I recognised them as the salt of the earth and committed myself to their care with complete assurance.

Our immediate objective was the broad saddle called Kabara that lies between Mounts Karisimbi and Mikeno, whose summits are about 4 miles apart. Climbing up to this saddle involves all the miseries and glories of high Africa, but as you ascend, one of the glories overrides all else. This is the hagenia forest. Hagenia trees grow more or less impressively on all the high mountains of eastern Africa, but nowhere, either in numbers or stature, so splendidly as here. An individual tree is a marvel: picturesquely gnarled and twisted like a giant bonsai, it can attain an immense girth, with wonderfully rufous, rugged bark. Its monstrously distorted limbs bear umbrellas of pretty pinnate leaves and carry a confusion of aerial gardens made up of moss cushions, hanging ferns, orchids and dangling vines. Such is a single tree, but here we had a great forest of them spilling out from the saddle and spreading up the mountain flanks. To complete my satisfaction, many of them were in bloom, seemingly weighed down with pendent reddish sprays of wistaria-like flowers. Massively bearded in flowing Spanish moss, and swathed in mist, this weird world seems to come from the pages of a fairy-tale illustrated by Arthur Rackham.

After about four hours we emerged on to the saddle at 10,000 feet, golden sunlight breaking through the cold, late-afternoon mist. The open space is welcome after the forest, and there is a pleasing sward of buffalo-grazed grass, buttercups, alchemilla and violets. At the edge of the meadow, where the slopes of Mount Mikeno start, I found a neglected corrugated iron hut, in which we settled for the night. I had read about this saddle, and knew that it had been immortalised as the burial place of Carl Akeley. Akeley was an American, and a visionary as far as the fate of the wild animals of Africa is concerned. He was the first person to realise how much the mountain gorillas were at risk, and also how great was their potential for research into pre-human anthropoid behaviour. It was through his intercession that the Belgians in 1925 created the Virunga Gorilla Sanctuary – the first national park in Africa – which in 1929 they greatly extended to become the Parc National d'Albert (now the Parc National des Virunga), covering much of the Rift valley from western Rwenzori in the north to Lake Kivu. Carl Akeley died at his beloved Kabara meadow in 1926 and was buried there by his wife Mary in what he had described as 'one of the loveliest and most tranquil spots in the world'.

Next morning, when the hagenias were dripping with the dense mist, I set out to go as high as I could on Mikeno, accompanied by the two elders. I was looking forward to their company; they both had such good wise faces, and such soft smiles and gentle manners that my heart was touched. Africa can produce such wonderful people: how privileged I have been, so often to have found myself in their company in beautiful and remote places.

We made no haste, covering the steep tangled ground in relaxed style, cutting as necessary with the *panga*. After an hour or two we left the forested slope and came out on to a narrow ridge amongst giant heath, and for the rest of our climb we stuck to this, with airy drops each side into deep ravines. There was plenty of evidence of gorillas in the form of their droppings, and Kokota pointed out to me their night-stop beds of twigs and leaves on the ground.

After another two hours, the mist clearing sufficiently, we saw the final tower of Mikeno at close quarters. It is formidably impressive, but in no way is it a climber's mountain. Its so-called rock is soft, crumbly and treacherous. Add to this that it is not far short of vertical, is heavily vegetated and usually wet and slippery, and you have good reasons for not lightly attempting the final summit. In spite of this, it has been climbed a number of times. But it was not for us this day. We settled for the point where our ridge abutted on the final face. My altimeter read 12,900 feet, so we were about 1,700 feet from the top. We rested for an hour in warm sunlight. The Rwandan countryside was a finely detailed patchwork quilt of intensive agriculture. By contrast the Zairean side was almost uninterrupted forest. What a monument to Carl Akeley!

Back at our Kabara camp, that evening I reflected on the dramas that had so curiously been played out at this remote spot. For more than thirty years after Akeley's death the gorillas had been safe in his sanctuary, which was rigorously sustained by the Belgian administration. In 1960, another American, George Schaller, set up camp at this spot for a prolonged gorilla study, during which he was to lay down the methodology for this kind of investigation. He revealed the social stability of the gorilla family group, and showed that it was possible for a human observer to become accepted within that group. What is more, he finally demolished the ignorant concept of gorillas being dangerously hostile to man. In his book, *The Year of the Gorilla*, he sums them up as 'eminently gentle and amiable creatures . . . . the dictum of peaceful co-existence is their way of life.'

Sadly, his was the last vision of this high-altitude Elysium in the heart of Africa, and the first premonition of the new era that has turned the region into a battleground.

This year, 1960, was the watershed year when the Pax Belgica, soon to be followed by the Pax Britannica, started to crumble, and Schaller's work was cut short by the descent of the Congo into chaos.

To their immense credit, the mountain gorilla story is very much a story of Americans. In 1967, another American, Dian Fossey (she was an occupational therapist working with handicapped children – perhaps an appropriate background for her strange life's work), under the guidance of Dr Louis Leakey, doyen of East African anthropologists, re-occupied Schaller's camp at Kabara with a view to continuing his studies. But within a few months she too was forced to abandon her plans and flee the country, a victim of the persisting anarchy. However Miss Fossey – she was 35 years old at the time, a 6-footer, full of doubts and misgivings but immensely gifted with moral courage – was not dismayed. Re-approaching the gorilla heartland from the Rwandan side, she set up a base on the equivalent of Kabara – on the saddle between Karisimbi and Bisoke that I have described earlier, only a mile or two from Kabara. Karisoke was to become internationally renowned for gorilla research, and here she worked for most of the time until her violent death in 1985, aged 53.

To start with, her preoccupation was with identifying the several gorilla family groups and habituating them to her presence as a precursor to recording their behaviour. She soon realised, however, that the gorillas' immediate survival was at stake because of threats from three different quarters – native hunters, cattle graziers and excisions from the sanctuary for cultivation by Rwanda's exploding human population. Her progressively bitter swing from gentle gorilla observer to militant gorilla protectress emerges with burning integrity from the pages of her personal testament,

Gorillas in the Mist, published only two years before her death, and from her diaries, posthumously edited by Farley Mowatt as *Woman in the Mists*. Her high-profile attempts to protect the gorillas made her enemies in high and low places. She suffered neither fools, nor those who differed from her, gladly. On the morning of 27 December 1985 she was found brutally murdered in her lonely cabin at Karisoke.

This tragic ending to one of the most remarkable stories in the history of wildlife conservation contains several lessons, not all of them obvious. At the simplistic level, the story can be regarded as a battle between good and evil. But there is another level which has to do with basic character flaws that have bedevilled almost all Western good intentions in Africa, and which I have recognised as much in myself as in others.

In Britain we pride ourselves on a slow, step-by-step process of change, based on experience, consensus and compromise. But time and again I have seen people who were brought up on this ethos at home, change overnight on coming to Africa, to become zealous missionaries condemning all about them, for whom it would be intolerable to compromise or make concession. Yet it has been the white man who has destroyed the wild animals of Africa. He has done this not just by his introduction of modern firearms and the concept of hunting as a sport, but also by new ideas of land usage and cash crop produc-

Below. *Dian Fossey's mountain gorilla study centre of Karisoke, on the 10,000 foot saddle between Mounts Karisimbi and Bisoke in Rwanda. The reddish wistaria-like flowers are hanging from a hagenia tree.*

tion. Far more profoundly, he has done it by removing the pre-existing natural constraints on human population increase and so upsetting the environmental balance and destroying the animals' ecosystems. If he had not come, the wild herds of Africa in the twentieth century might have remained as glorious as they were in the previous millennia, when man and wild creation lived in balance and the word 'poacher' had not been invented.

The white man made the African a poacher on his own heath by the simple process of declaring that the hunting which had been an integral part of his lifestyle since time immemorial was suddenly to be illegal – while the white stranger arrogated to himself a privileged position for sport, or massive game desruction for such questionable objectives as grazing protection or tsetse fly control. Readers may be incredulous, but the quite appalling fact is that over vast areas of eastern and central Africa mass total large game animal eradication was carried out for the purpose of removing the natural hosts of the tsetse fly, which otherwise prevented the introduction of the ranching of domestic cattle in such areas, because it infected them with *nagana* – trypanosomiasis – a similar infection to sleeping sickness in humans.

Is it a matter for surprise that when the white man abdicated his overlordship a quarter of a century ago, the Africans once more took up their birthright? But by now the potential for abuse was enormously enhanced: modern weapons and transport made large-scale hunting, for motives increasingly removed from those of the original tribal hunters, practicable for a new class of profit-motivated people.

Returning to the gorillas, there are organisations in the Western world that will offer five figure prices (in dollars) for baby gorillas. But to catch a baby gorilla, at least the mother and probably other members of its family must first be killed. With such sums on offer, it is not surprising that corruption runs all the way down the line from foreign embassy and wildlife ministry, to local police and park wardens, and to the ignorant and minimally rewarded hunters. Character degradation is integral to the international trade in wild animals: unfortunately it can lead to an evangelical over-reaction from those who see themselves as being on the side of the angels. They feel they will be judged by the overt vigour of their response: armed patrols and game scouts, ambushes, Landrovers, helicopters – such are the lines on which the Western mind runs. Macho measures appeal to minds brought up on the spirit of the way the West was won. They may have a part to play, but when one looks at the history of Africa over the last century, it is clear that they have failed and a radically new approach is needed.

These thoughts bring me back to Miss Fossey. That she devoted precious years to the development of gorilla habituation is the message from her book. But from my own experience of admittedly different situations, I wonder whether she might not have done better to leave the gorillas to themselves and to attempt to habituate the poachers? Such men are not essentially malicious, nor do they hunt for self-gratification as Europeans do. They hunt – whether gorillas or humble forest buck – in response to economic pressures. At its crudest, gorilla conservation funds might have been spent more effectively on buying off (or, to put it more respectably, compensating) poachers for loss of earnings. Lord knows, paying people for not doing things has become a major feature of conservation in the West. Such an approach might have been more cost-effective than conducting a private anti-poaching war, and this could have led to a more permanently effective state of affairs whereby the poachers themselves were made gamekeepers.

I can believe that this would have been much more difficult and much less fun than habituating the gorillas, and would have called for different skills. Indeed, from what emerges of Miss Fossey's character, she would have been quite unsuited to it. Anthropologists, sociologists, teachers, linguists – people of this mix are perhaps more important in the battle to save what is left of Africa's wildlife than zoologists, ecologists, policemen – or veterinarians such as myself. Tact, subtlety and guile may be more potent than taking up arms against a sea of trouble.

It was with these thoughts in mind that I set out to visit the mountain gorillas of Zaire. These creatures are the largest of the apes, and they are a separate sub-species from the far more numerous lowland gorillas of western Africa. They were discovered on Mount Sabinyo in 1902 by an officer of the German administration, Captain Oscar von Beringe, and so they bear his name – *Gorilla gorilla beringei*. Never very numerous, their population was much reduced by white hunters in the first quarter of this century, while over the last twenty-five years they have suffered even more disastrously from the drastic reduction of their rangelands. They are now an endangered species, probably numbering less than 250 individuals in the volcanoes and perhaps 150 elsewhere, whose survival must be questionable in spite of international efforts to save them. This is what *Homo sapiens* has done to one of its closest evolutionary relations, in less than a century.

Although Mount Mikeno and its Kabara meadow are historically so closely identified with mountain gorilla studies, the Zaireans have based their study project at Jomba, some 6 miles to the north-east, towards Mount Sabyinyo. Here the mountain fan-slopes are more extensive and, having been largely protected by the national park, they provide a greater area of the lower-altitude vegetation that suits the gorillas. My driver Abdallah and I took the road northwards past Kibumba, through the little centre of Rumangabo, about 28 miles form Goma. Here we turned to the east, towards the mountains, and at once found ourselves on a totally execrable track which eventually just petered out. We were advised by a woman who was hard at work with her children, baby strapped to her back, weeding

coffee, that from now on we had to foot it up a path amongst gardens and pastures. Happily it was a simply beautiful day, bright and fresh, and porters having been found, I enjoyed the walk upwards towards the dark forest.

The project's headquarters are well sited with airy views and there is a pleasant little wooden guest house. Later in the afternoon, as I sat outside on the earth terrace amongst sweet-flowering masses of cosmos, I saw a small safari of porters winding up the path from below and soon found myself joined without pre-arrangement by two Europeans. One of these was Professor Ernst Lang, a Swiss veterinarian and authority on primates, who has made outstanding contributions to gorilla studies, and the other was Dr Wolfgang von Richter, a German who is Technical Counsellor for the Kahuzi-Biega gorilla sanctuary in the low mountain forest at the southern end of Lake Kivu. Thus for my visit to the Zairean gorillas I found myself in the company of two of the best-informed companions one could wish for.

We sat up late that night, looking out over the vast moonlit expanse of the Rift valley to the west, with its tiny pinpricks of light here and there indicating the home fires of the Zaireans, and we thrashed out the problems of forest and wildlife conservation in Africa. It was some consolation that all three of us, through our long and very different professional experience, had arrived at the same view, whereby we put human population containment and ecosystem rehabilitation at the top of our list of priorities, ahead of the conventional technologies to which the great preponderance of Western aid has so far been devoted.

We were up at first light next day, when the volcanoes were still crisply silhouetted against the primrose, pre-dawn light in the east, and set forth with our tracker Ruwelenga and his mate, both dressed in faded and much patched green denims. Soon we were immersed in the dense vegetation of the mountain fan-slope. For about an hour, in the cool, moist, early-morning air, we quietly pursued our way along narrow tracks over rising country, warming to our work as the sun rose higher in the silver-flocked sky. As we progressed, the going became more difficult: there was now no pre-existing track, just here and there the brushed effect on the vegetation caused by gorillas passing before us in extended formation. There followed another hour of starting and stopping, and a familiar feeling of despair arose in my heart: surely it was all going to prove fruitless.

Our Zaireans crouched; we did the same. They communicated, apparently inaudibly; we did the same. They moved on a few paces, repeating the process, silently slicing a few vines and nettles with their *pangas* to ease our passage. Much of the time we were on our hands and knees, and at ground level the air was cool, smelling of dank humus. Bamboo and mimulopsis crowded over our heads – visibility was no more that 3 or 4 yards and quite hopeless for photography, my faint heart told me. We continued like this for another twenty minutes: at least we were not being exhausted by excessive physical demands, I thought thankfully.

The trackers yet again shrank to the ground and froze, and we followed suit. After half a minute Ruwelenga looked round quizzically, catching my eye as if to say, 'Well, what more do you want?' – and then looked ahead at a dark tunnel of vines and nettles. As my eyes adapted, I realised that I was looking at a huge expanse of badger-like black and silver hair – a male gorilla, back towards us, immobile. As we watched, an arm of seemingly great length and flexibility twisted round backwards and a large black hand, like someone wearing inflated black rubber gloves, began rythmically scratching the silver spine!

His lack of concern at our presence was positively disconcerting. Perhaps he read our thoughts, for after a minute he silently pressed his hands knuckle-wise on the ground, slowly turned round, thrust his head forward and, from a distance of 4 yards, inspected us.

I confess that I had set out on this gorilla chase in a slightly blasé frame of mind. In my life I had had the good fortune to see almost all the major species of wild animals of eastern Africa in their natural haunts: why this mystique about gorillas? But in this magical moment all such arrogance was demolished and my reservations were swept aside. He looked at me, it seemed, at the least from an attitude of equality, if not indifference, and the face, with its deep-set black eyes and great domed forehead, was sentient in a different way from almost all other wild animals. Here was a gravity that set him apart from his only serious African rival, the chimpanzee: for the chimps, just as close to us genetically as the gorillas, have a slightly comic pathos about their looks. But there is nothing either comic or pathetic about the gorilla.

The subject we were now facing, eye to eye, was a mature male whose weight we estimated at 350–400 pounds. This is not particularly heavy for a gorilla male, but nevertheless it was between two and three times my own weight. We had already seen his handsome silvery back, as though dusted with hoar frost from neck to loin, but now we could take in his massive cranium with its great sagittal crest and his shiny, black-leather, prognathous face, dominated by its large, flattened, backward-sloping nostrils. It was the complex etchings and folds of this nose that gave Schaller and Fossey their individual imprint system of identification – the gorillas' nose prints.

He was squatting complacently, his short legs invisible, just his feet and toes protruding below a rounded Buddha-like belly, while his proportionately longer arms, with their great hands and fingers, wandered casually about in the greenery picking here and there a leaf to nibble. We were accustomed by now to the greenish light and, at last overcoming our feelings of awe, we set to work with our portrait lenses – no use for a telephoto here. I was squatting, using my camera over the shoulder of our tracker. As I framed the animal for the fourth or fifth time, he casually rose on his legs, pressing his body up into near-6-foot stance with his long arms. He stood there, 4 yards away in his green tunnel, and then, in a split-second, gave forth a chilling shriek – the best word I can think of – and launched himself bodily forward. He landed with outstretched arms at the feet of Ruwelenga, his great black hand grasping the tracker's forearm, his eyes staring into his face from a distance of 10 inches. He had landed, incidentally, within 2 feet of where I was squatting and I felt the ground shake and could smell his horsy-sweet body odour.

Below. *A young member of the mountain gorilla family group amongst whom we spent an enchanting morning on the Zairean flanks of the volcanoes.*

My knowledgeable companions had drilled me well in gorilla protocol. One crouched, one bowed one's head submissively, one kept motionless and silent. On no account did one raise an arm or point, let alone turn or retreat. Ruwelenga knew the game: this was not an attack but rather a testing of relationship and nerve, and a reinforcing of the gorilla's sovereignty over his family group. The tracker waited a few seconds and then with gentle sang-froid slowly detached the gorilla's fingers one by one from his arm and returned them to their owner, who thereupon edged himself backwards to his previous stance and once more took up leaf nibbling.

We had only just recovered our composure and returned to our cameras, when the whole mock attack was identically repeated; but evidently we passed the test, for the gorilla turned unconcernedly and vanished into the green tangle.

Now from all around us we heard a grunting and coughing and muttering, and the thrilling puk-puk-puk of chest-beating. All this was in effect the intercommunication of a gorilla group, telling of the advent of a human group. Our trackers knew what they were about: moving bent down in the vegetation, a few paces at a time, we emerged into a more open area, about 30 yards across, where the bamboo gave way to tussocky grass, thistles, umbellifers and vine-draped hypericum trees, and there we found ourselves surrounded by the whole gorilla group. Here was our patriarch, several younger males yet to sport silver capes, mothers with babies at their breast, and wickedly independent youngsters. These last were

Right. *The patriarch silver back of the mountain gorilla group which I visited in Zaire. We estimated that he weighed about 450 pounds. Within a few seconds of taking this picture in the dense mimulopsis tangle, he stood up vertically and threw himself at my tracker, seizing him by the arm. Note his wrinkled 'nose print' and the huge saggital crest on his head.*

the only ones who climbed the rather slender trees: for the greater part, gorillas are terrestrial. The young tree-climbers, however, stared down at us from their look-outs, hauling in lengths of the galium vine, a favourite food which they ate by pulling lengths through their mouths, and deleafing it with prehensile lips and tongue.

When we first emerged into the glade, the gorillas (my all-up count reached thirteen, including two babies) were evidently nervous and curious, trying out little acts of daring on

totally abandoned way; but now and then one or two would make a serious effort to take a snack, either from the vines or the umbelliferous wild celery which here grows as a 6-foot-tall plant with crunchy, watery stems. Meanwhile the young ones larked about, taking increasing liberties with their elders until slapped for their cheek.

The mothers are only about half the size of the adult males and they feed their babies at two rather diminutive breasts high on their chests. I spent some time watching a mother who had tucked herself into a hollow of vegetation at the foot of a St John's wort tree. I guessd her baby weighed about 10 pounds: gorillas weigh about 2 pounds at birth, and this baby was possibly about 6 months old. It would be an understatement to say it was hyperactive: it did everything possible to attract attention, clawing its way over the mother's coat, peering into her eyes, jumping over her, biting her toes, rolling off, turning somersaults. The mother remained patient and unruffled throughout. The young one then discovered a hanging loop of vine and started swinging to and fro on it; then he jumped from the swing on to his mother's chest, grasped a breast, took the nipple into his mouth and, clutching her hair, remained sucking, totally immobile.

Gorilla babies have a more prolonged period of dependence on their mothers than any other kind of animal. They breast feed for over three years, and while this continues the mother's sexual cycle is suppressed; hence the minimum breeding interval is not less than four years. Thus even totally protected populations with unrestricted dietary range can only build up very slowly. However, this may be to some

us; but soon the tension seemed to lessen and they got on with the business of the day, indifferent to our presence. That business, now that we were in dappled sunlight and the vegetation was drying out, seemed to be largely sunbathing and loafing – lolling about in a extent offset by the fact that their probable natural lifespan is 40 to 50 years, and as far as is known, they breed throughout this time.

So we passed the morning hours in this Garden of Eden until, one by one, the gorillas disappeared and Ruwelenga signalled that it was time for us to retrace our steps. We left as silently as we had come: this was not our domain, and we should leave it discreetly, with proper humility.

On a different occasion I made a similar visit to the gorillas of Mount Bisoke in Rwanda. Here the procedure is controlled by the Parc National des Volcans authority, under the guidance of the Mountain Gorilla project. One of their objectives is to obtain funds for conservation by encouraging 'gorilla tourism'. They have been so successful that they are in danger of suffering from too many visitors. The Gorilla Project staff consider that one visit per day per gorilla group, by a maximum of six visitors, is as much as the creatures should be required to tolerate, for fear of causing stress. Since tour and study groups from Europe and America increasingly make block advance bookings, the individual traveller must book far ahead – or use his wits! This degree of organisation and the fact that the gorillas' range in Rwanda has been so severely restricted by agriculture, makes a visit an easier but perhaps less enchanting experience than in Zaire.

None the less, for people of the right turn of mind, I don't hesitate to recommend the attempt. There is something special about these animals, and it is not possible to be in their presence without philosophising about our own race and our relative place in the scheme of things. As you peer through the tangle, intruding upon your fellow primates, surely the spirit of Charles Darwin is leaning over your shoulder. This is to exclude, at least in my case, any anthropomorphic element, although I know others take pleasure in this – and why not? But vets are seldom anthropomorphists, in the sentimental sense. You will seldom hear a vet

say that he loves animals or sees human qualities in them, but they interest him absorbingly and fill him with a sense of compassion. This is quite enough for me, and amply justifies the effort required to find these gorillas in their sadly diminishing mountain forests in the middle of Africa. Is it stretching things too far to suggest that there is something appropriate in the idea that the birthplace of the Nile – the river that has been the cradle of mankind – should be occupied by this species that narrowly missed out in the race to dominate the earth?

Opposite. Peltigera canina & Marchantia polymorpha. *The complex ground flora of the middle altitude belt of Rwenzori forms a unique raised wetland that serves as a vast reservoir of the Nile. I asked Christabel King for a painting to signify the importance of this biosystem of lowly plant species and she produced this life-size study of a strangely beautiful* Peltigera *lichen and a* Marchantia *liverwort, both with their fruiting bodies. I found them in abundance between the Kitandara lakes at 13,200 feet.*

Peltigera canina

Marchantia polymorpha

RWENZORI FROM END TO END

In 1984 my plans for a journey up the Nyamu-gasani valley from the southern end of Rwen-zori had been frustrated by the Rwenzururu disturbances. Instead, I had entered the range from the south-east, ascending the Nyamwam-ba valley from Kilembe, and so visiting the Nyamugasani chain of lakes from the upper end of the valley. Now, in July 1988, I deter-mined to make a last attempt to enter the mountains by the lower Nyamugasani valley, hoping that the disturbances would have quietened down, and we assembled rations for twenty-two men for eighteen days, to allow a safe margin. The District Administrator at Kasese gave me the all-important letter of authority, but with the caution that we should none the less obtain clearance from the Nation-al Resistance Committee at Kyarumba in the south. At this, Old John Matte shook his head dubiously, and kindly insisted on accompany-ing us to our point of departure.

The arrival of our lorry at the road's end at Kyarumba caused a sensation and we were soon surrounded by a multitude of several hundred people. After the hassle of packing up at Ibanda and the wearisome road journey, I was tired out, but any hope that we might be allowed to set up camp straight away was dashed by the arrival of an officious messenger: I was required to attend at once at the office of the 'NRC 1' – that is, the most lowly of the administrative and political units that make up President Museveni's system of government. My party were impatient at this bureaucratic obstruction: they had brought with them their Nyabitaba staves, being uncertain whether they would be able to cut them on this unknown route, and now, to a man, they fell in behind me to form a Praetorian Guard. The multitude, swelling from moment to moment, followed on and in this fashion we wended our way through the streets of this large village until we arrived at a small, open-fronted shop.

Here, amongst the soap and matches, sauce-pans and baskets of flour and beans, a table had been set up, on the far side of which five sombre, Cromwellian-faced Africans were sea-ted. These constituted a quorum of the village Resistance Committee; I gravely shook hands with each in turn.

My letter from Kasese was regarded only as a starting point and I had to endure a laborious interrogation as to my background, opinions and motives. John, Moses and I were crammed inside the shop, but outside the street was now jam-packed with people. However, the mem-bers of my Praetorian Guard had all worked their way up to the front, just outside the shop, and established themselves as a rent-a-crowd. Each question, answered first by me with modesty and discretion, was then answered with embarrassing embellishment by John. 'Certainly a man of enormous probity . . . who is more liberal than anyone when it comes to his employees . . . Why, he has devoted his life to these mountains and knows them better than any of us.' After each such remark, John turned to face our men and asked, 'Isn't that so?' – whereupon, with their staves simul-taneously raised to make a small forest, a throaty roar in Rukonjo responded, 'The sooth indeed – he speaks nothing but the truth.'

This impressive display of solidarity won the day. Paper and ink were sent for, and authority was written out and rubber-stamped, and the Committee rose with smiling faces, shook me by the hand, wished me well for my journey – and then, to my absolute disgust, announced that the messenger would take me to be interviewed by NRC 3! (NRC 2, the Parish Committee, thankfully was not considered to be involved.) The concourse of people followed us the several hundred yards to an apparently identical shop where the procedure was repe-ated, the only difference being that my private army was now, like me, seriously tea-deficient and so even more impatiently interventionist. Another rubber-stamped authority having been pinned to the previous two, I was told that I was to be interviewed by the Gombolola Chief. This person is a government appointee of quite senior rank, his writ extending to all the rural areas around Kyarumba. This was to be a private meeting in his white-washed, mud-brick office, my rent-a-crowd none the less taking up an ostentatious position at the entrance to his compound. The Chief was a charming man in his thirties and we enjoyed a free-ranging conversation. On my turning this to the matter of over-population I found him receptive, and he was expansive on the urgent need for nation-wide family planning.

While we were talking I noticed that numbers of small, smiling faces were eagerly peering into the room to view the stranger. I had just asked him if he had any family: as always with Africans, his face lighted up.

'Oh yes,' he said, 'of course,' and waving his hand in the youngsters' direction, he said proudly, 'these are my children.'

In astonishment I made a careful head count – there were *twelve*! 'What!' I said, 'are these *all* your children?'

His face lighted up again, 'Oh no! – not quite all – the older four have not come back from school yet!'

He kindly placed the pleasant grassy area behind his house at our disposal, arranged water, firewood and gifts of vegetables and fruit, and at last John was able to take his leave and we were freed to attend to our domestic affairs.

Kyarumba did not let us go lightly. Next morning, as our column of men filed up the path above the village, we were accompanied by almost all the inhabitants, making a long, dark, winding snake amongst which here and there I could indentify my porters by their loads and staves. We crossed the raging Nyamugasani torrent by a springy bridge of two or three eucalyptus trunks, here leaving behind most of our adult well-wishers, but well

over a hundred children continued to accompany us with joyful voices, and it was thus, Pied Piper-style, that I set out for the Mountains of the Moon up the Nyamugasani valley.

For a local guide we had been advised to try to find a certain old man, Matayo Muhindo, of whom we were told in quaintly Wenceslas terms, that he lived at Kabingo, 'hard against the forest edge', at the limits of human habitation. It took us the whole tiring day, halloaing up and down hillsides, to find him. I had been told that he was old, but I had not expected him to be quite so critically – one might almost say terminally – old! But for all his wasted form he was lithe, with a stalking gait like a lively bird, while his black eyes sparkled below his grizzled balding poll.

He joined us in camp that evening at Mubalya, equipped for the long and arduous journey that lay ahead: that is to say, as well as his ragged trousers and coat, he carried a home-made bush knife and a small monkey-skin pouch containing a pipe and tobacco. We had, however, anticipated his needs, and when Moses gave him a blanket and pullover he was quite overcome.

Our route for the next four days took us up a complex ridge to the west of Nyamugasani, ascending some 6,000 feet, with the marvellous castellated mass of Mount Rwatamugufu, constant birthplace of the clouds, across the valley to the east. The name has been translated as 'The Strong Man' or, more picturesquely, 'He that breaketh your bones'. There was no trace of previous human passage – only numerous punch holes in the earth, showing that we were following one of the timeless elephant walkways of Africa; but Matayo said that no elephants had been here for the past ten years.

By the time we were approaching our sixth camp the weather, which had been pleasantly summer-like for the first few days, had deteriorated into almost permanent Scotch mist. Having at last broken out of the dreadful heath-moss belt at about 12,000 feet, we crossed a rocky stream flowing through a pleasant alchemilla meadow and came to a fine rock shelter under a giant erratic. It was large enough to hold our party with ease, but my presumption that we should camp here was dismissed by Old Matayo. He pointed to a formidable series of rock terraces to the north, and insisted that we had to climb over them to a pass, where we should find a place to camp. On putting the question to the men, I was outvoted: it proved

Below. *Mzee Matayo Muhindo, the delightful old Konjo gentleman whom we co-opted to be our guide for our 18-day journey through south-eastern Rwenzori in 1988. The bleb-like scars on his forehead – beautifying marks – show him to be a child of an earlier era. We worked out that he was probably about 80 years old – 12 years my senior!*

to be an unfortunate decision for us.

In drizzling icy rain we set about the beetling cliffs and terraces of rock and moss, an ascent not far from vertical of something like a thousand feet. My party closed up, so that one man's bare feet were level with the following man's head, the second man often grasping the first man's foot to help find a foothold. It was absurdly dangerous for men so bulkily laden, and it was all I could do, speechless with fatigue, to follow them.

Dark clouds engulfed us as we reached the crux in the face of driving, bitterly cold sleet, when whistle signals came back from our scouts warning us not to proceed: there was no possibility of a campsite, and now we were within the last hour before darkness. We held a huddled council, and testily, I insisted on going back down the cliff and camping at its foot. No one dared to dissent, and miserably cold, wet and despondent, the column of men slowly disappeared over the lip, I myself waiting till last to make sure everyone was gathered in.

In the failing light, when we were less than a quarter of the way down the cliff, out of the mist I heard a whistle that I knew meant 'come this way'. Instead of descending further, my men had followed a narrow horizontal traverse just wide enough for their feet, which led to an uneven rock platform 5 or 6 feet wide, sufficiently overhung to provide some dry areas. Beyond this a curtain of drips, raining down from above disappeared into the frightening depths below. Avoiding further descent overrode all other considerations to me, and I readily agreed to bivouac, however uncomfortable the prospects were for the night. One part of the shelf extended a foot or two further outwards, catching the drip line full pelt, and on to this my igloo was squeezed.

Within minutes of my approval, a helichrysum fire was blazing and soon massive stems of groundsel were being manipulated down the cliff face on to our shelf. As the sun set, the cloud to the south lifted and we found ourselves with a fantastic view right down the

Nyamugasani valley to the still-sunlit plains of Uganda beyond, with Lake Edward, a silver sea, filling the southern horizon.

Because of the narrowness of the shelf, the men made several fires, for which the out-curved rock face acted like the back of a giant grate as the smoke and sparks soared upwards. Clinging to the cliff in the darkness, their black, hunched forms silhouetted against the rock-face, illuminated in the flickering firelight, my party made a thrilling spectacle. They could have been a band of vagabonds or *banditti* from a picaresque novel.

Two days later, still in thick mist, we were once more trapped in the heath-moss tangle, and found ourselves traversing horizontally the increasingly steep side of one of Rwenzori's characteristic U-shaped valleys. Alarmed that we were getting ourselves into a dangerous situation, I had just decided to call a halt when the porter in front of me – a steady lad called Reuben Asimwe – suddenly shot down the hillside with alarming velocity, somersaulting twice before managing to stop himself by grabbing some helichrysum. His load went careering on but fortunately jammed in a twisted heath-tree root. His Konjo brothers guffawed at his discomfiture, but I was almost as shaken as he was and consoled him as, happily unharmed, he struggled to recover his load. I think this is the only time I have known a Konjo to fall seriously.

Below. *Porters struggling across carex tussock bog in the upper Nyamugasani valley, south of Lake Katunda. Their heavy loads tend to pull their heads backwards, and they often try to counter the thrust with their free hand.*

My mind made up, I passed the message to recall our trackers, but on their return, to my dismay the consensus was that Reuben had inadvertently shown us the best way, and our column turned straight down the valley side. The going was desperately steep but after an hour I heard the kind of whooping high-pitched calls that I knew meant 'Its OK – at last we've got somewhere.' The gradient lessened, the tangle gave way to grassy tussock and to my relief we found ourselves on a fairly level alluvial fan beside a meandering stream. Now the mist started to lift and revealed that we had descended the west wall of a quite amazing circular rock amphitheatre. It was about a quarter of a mile in diameter, with cliffs of 500 to 1,000 feet high, except where to the south-west the stream made its exit by a dramatically spectacular gorge.

We at once decided to camp, and round our fire I quizzed Old Matayo, but he denied any knowledge of this extraordinary place. Its remote inviolability, unvisited by humankind, even seemed to cause wonder in the minds of the normally prosaic Konjo, who are not given to marvelling about their surroundings.

On the following day we reached the great rock shelter of Kinyamuyeye – one of the largest and best in the mountains – and for the first time in eight days I knew exactly where we were. We shortly joined my route of 1984, and after camps at Lake Katunda and Lake Kanganyika, struck east to make a new route back to Ibanda, camping in dense mist on dry tussock on the eastern slopes above the Nyamugasani.

We had now entered country that none of our party had ever traversed before – not even Kajangwa, on whom we so heavily relied. Next morning, in soaking Scotch mist that was to persist without let-up for the next three days, we moved off on an easterly bearing. All my prior experience in Rwenzori had been based on either valley or ridge routes, on which you can usually make a reasonable shot at a map fix by altimeter reading. But now we found

ourselves committed to *terra incognita* by a route that cut across the grain of the country, passing up and down over innumerable ridges and valleys.

After a few hours, our sense of disorientation was complete. We were prisoners of the intricate topography and this rendered my compass and altimeter almost useless. We were in a totally silent world in which not a breath of movement occurred in the saturated air, and there was no way of telling, until within a few paces, what awful obstacle would next present itself. A frightening drop into nothingness might open up at our feet, or a hardly ascendable cliff of moss-covered rock, or a bog of unknowable depth and breadth, where I feared my party might become irretrievably dispersed. Massive rock moraines carried some of the finest stands of heath forest that I had ever seen: doubtless sublimely beautiful given just a touch of sunshine, they were truly awful – even frightening – in our desperate circumstances.

Late that afternoon we found ourselves at the foot of a smooth rock face whose extent neither upwards nor lengthwise could be determined in the mist. It had just the slightest overhang, so that there was a terrace of dry ground a yard wide along its base, and here, tired and dispirited, we bivouacked, the porters lying head to toe along the narrow belt. So miserable was the persisting rain next morning and so uncertain the path ahead that it was with unusual difficulty that I persuaded my party to leave their fires and set off. This day's march was even worse than the previous one. We had entered the great glacier-scraped wilderness of Kianamo – a labyrinth of boggy stream beds, heath-forested terraces and islands, and unnervingly steep ravines. This was the ultimate in virgin primeval Rwenzori: in clear weather it could have been a delectable experience; as it was, it was fearsome and soul-destroying. The only reward of the day was the discovery of a dramatic fall of white water, cascading from mist-concealed heights, spilling in silver ribbons over the black rocks and disappearing into the misty depths below. There was something rather eerie – even sinister – in this display of immense power coming from nowhere, going nowhere.

Our bivouac was as wretched as the night before and when next morning we found the mist as dense as ever, alarm bells began to ring in my mind. Everyone agreed now that we were totally lost, and no one had any suggestion as to our best course. How much longer could we afford to make no tangible progress? We had set out with a safe margin of food, but were now days behind schedule. In the grey obscurity of these powerfully intimidating realms of mist, I began to feel a sense of impending disaster. I realised that I was too much in the habit of total dependence on the Konjo to solve route-finding problems: now, as happens with migrating birds caught in the winter fogs of Europe, they were disorientated, demoralised and silenced.

It was a rare moment for me to exert authority. Compass in hand, I joined Moses, Kajangwa and Matayo in the van, and as rigourously as I could, for the rest of the day insisted on a northerly bearing. To stick to a compass bearing through dangerously broken country in nil visibility is a recipe for every kind of difficulty, but it was better than any softer option. Not just myself, but the others too, were constantly slipping and stumbling, but proceeding thus painfully slowly, after some hours Moses suddenly said 'I can smell smoke,' – and the others halted, sniffing the still air like retrievers. Rounding a rock bluff, we discovered a fine rock shelter, where we were first greeted by a barking dog, next by a sweet-faced boy of perhaps 10 years old, and finally by a little grizzled old man who was sitting watching over a cooking pot set on a fire of glowing ashes.

I shook him warmly by the hands, knowing that with this chance encounter our troubles were over – he would give precise instructions to my trackers. Meanwhile the rest of my party all crowded into the cave, their spirits visibly

soaring: the fire was made up and large-scale tea-making operations set in train while our host, bemused but delighted to have unexpected company, explained that he and his grandson were here on a hyrax hunting trip. The evidence of this was all about us in the form of snares and twine, hyrax skins drying on a line and skinned 'oven-ready' hyraxes being smoke-cured in the cavern roof. In answer to Moses' urgent questions, he told us to our astonishment that we had overshot both Mount Rugendwara and the Kuruguta valley and were now on a hunter's trail leading to Lake Mahoma – far to the east of our intended route, but well placed for a return to Ibanda via Nyabitaba. This bit of luck gave my companions enormous respect for my compass!

As we squatted there making tea, Moses bade me look round. The mountain spirits, having had their sport with us but now finding themselves rumbled, withdrew the veils of mist, revealing forested ridges tumbling down into valleys filled with silver cloud far below us, and there, in the distance like a bright new sixpence, was the tiny hill-top lake of Mahoma. I took a compass bearing and by back-plotting to the altimeter reading, was able to fix our position exactly. We still had a long way to go to the lake, but with restored confidence and high spirits, I knew that we should have no difficulty in eventually reaching Ibanda.

For so much of the time in these mountains one is enveloped in silent mists, and this can indeed lead to a depression of the spirits. But sooner or later there comes a moment when you sense that the mist is becoming mobile and

Overleaf. *At its southern end, Lake Kanganyika (which lies at 12,500 feet in the upper Nyamugasani valley) narrows before cascading down to join Lake Katunda. I camped at this mysteriously beautiful spot in 1988 and took this photograph about mid-day in dense motionless mist. Air and water are indistinguishable, except by the mirror images of giant heath.*

luminous, and then the sun breaks through, bringing a transformation that uplifts the heart and mind. The sudden brilliance of the world of golden mossy colours is ineffable, while from every tiny, dew-laden point, gem-like sunlight is refracted. At once all is forgiven and one's heart soars in gratitude.

The Konjo, who with their dour stoicism seem to have little sense of the unusual beauty of their domain, none the less reveal their souls by the way their faces change, their voices lighten, and song and laughter break out. The word animist is sometimes used in a depreciating manner by people whose minds are constrained by formal religion, and this word has been used by anthropologists in respect of the Konjo. But to me it simply expresses a state in which one is free to sense and to respond without inhibition to the spirit of special places and occasions which somehow have a quality of resonance with the fears and joys that struggle continually for the mind of man.

Thus far I have described Ugandan routes into Rwenzori from the east and south. Before following my northern journey, I will briefly mention the western approach from Zaire. On this side, because of the tilted nature of the range, gradients tend to be steeper, and only one route from the Semliki valley to the central peaks has been consistently used. This is the route I took at the end of January 1987, starting at the headquarters of the Park National des Virunga at Mutsora. I took a guide and four porters, determined on a slow ascent with two nights at each camp, to allow plenty of time for acclimatisation and photography.

The route runs at first on the northern side of

Opposite. *In August 1988 with a party of 21 Konjo, I was totally lost for over three days in the Kianamo region of south-eastern Rwenzori. The ambience of weirdly shaped giant heath trees in the silent mist can induce a sombre frame of mind. In the picture, our path-seekers are carrying not spears, but staves and* pangas.

the Butahu valley as far as the first hut at Kalongi (6,700 feet) and then, after crossing the Kanyamwamba, follows the ridge between that stream and the Kamusoso to the Mahungu hut at 10,900 feet. This is the original route of the German explorer Franz Stuhlmann in 1891, and his highest point, at 13,200 feet, is to this day called Campi ya Chupa (Bottle Camp) because he left a record of his climb in a bottle at this point. The Belgians built their third hut high up on a spectacular knoll at Kiondo, 14,100 feet, and this last stretch provides fine views of the snows and glaciers of the west wall of Mount Stanley, and also of Lac Noir lying sombrely in a dark glacial trench below.

All my African companions having pleaded illness of one sort or another, I made a solo ascent of Wasumaweso and was on the summit ridge before eight o'clock in the morning to find the whole range cloudless. My vista was against the light, looking east to where the sun, still below the high mountain horizon, was creating a dead-pan light of crisp clarity. But in due course it broke the skyline and progressively illuminated my side of the peaks, first obliquely and then increasingly face on.

This is a spectacularly dramatic mountain wall, providing a 2,000-foot vertical sweep of ice falls and hanging glaciers, corniced ridges and fluted snow slopes from Point Albert in the north to Savoia, Elizabeth and Philip in the south. Alexandra dominates the scene, a classic pyramid of black rock, ice and snow.

Below, Lacs Noir and Vert, so aptly named, were lying in their deep troughs, while the tiny tarns, Lacs Gris and Blanc, were tucked in at the foot of the Alexandra ice fall. To the west, the whole Semliki valley, the western Rift and its far blue escarpment walls, told me of the heart of Africa sloping to the Atlantic Ocean. I could scarcely credit my good fortune. So often in Rwenzori the cloud has robbed me of any reward for the all-consuming labours she demands: now, just this once, she was showering her bounty on me with both hands.

Next day, with signs of the fine spell breaking, I persuaded one of the men that he was well enough, and we made the steep traverse above Lac Vert to the foot of the great ice fall of the Alexandra glacier. The greenness of Lac Vert is astonishing, presumably the effect of algae in the water – the wind ripples, the reflected sunshine on its surface, and its picturesquely indented shape make it outstandingly attractive, especially where obelisks of Wollaston's immensely tall lobelias stand statuesquely on the slopes above.

This Zairean approach to the Mountains of the Moon lacks some of the mystery and challenge of the Ugandan routes, but in its way it is perfect, and for those whose aim is quickly to obtain close views of the snowy sources of the Nile, it is to be recommended.

After ascending the Nyamwamba valley from Kilembe in 1984 and exploring the upper Nyamugasani valley, I had crossed the Bamwanjara pass and then, via the Kachope and Kitandara lakes, crossed the Scott Elliot pass and descended to Bigo. From here I made a visit to the remote Lac de la Lune which lies over the pass at the head of the Mugusu, whose stream joins the Bujuku in the Bigo bog. Although in its middle passages the Mugusu contains extensive areas of dense giant groundsel forest which are difficult to penetrate, this valley is attractive and thrillingly enclosed by high mountains. Towards the end of August in 1984, in the company of Peter and two porters, I reached the Roccati pass (named after the Duke's geologist), which is the 13,500-foot col between Mount Emin to the west, and Gessi to the east. Just short of the crux in the lowest buttresses of Gessi there is a large rock shelter within which I was able to pitch my tent.

A few paces over the pass yielded a dramatic revelation – Lac de la Lune, remote, mysterious and rather forbidding as cloud rose up the valley from the north. Its walls are immensely steep, giving an impression of great depth, and it is entirely enclosed, its outlet being subterranean. This is the source of the Ruanoli river which runs north to the Semliki. These features, together with its position squeezed tightly in between Emin and Gessi, led Noel Humphries, fancifully to propose it as Herodotus' bottomless lake between two peaks, Crophi and Mophi – the ultimate source of the Nile. Certainly on evocative and aesthetic grounds it would do very well for this – it is indeed a veritable moon lake.

The world beyond our cavern was white with snow when we awoke next morning, and Peter and I set out unenthusiastically in dense weather to try to gain the main ridge of Gessi. The day turned into a desperate 2,000-foot struggle up a steep gully filled with helichrysum and groundsels, and slippery, moss-covered rocks. By midday the vegetation began to fade out and we climbed over snow and a few unconvincing remnants of what presumably once had been the Gessi glacier, to reach the ridge in white-out conditions. As we turned north to the Bottego summit, the ridge presented a series of giant rock blocks at which Peter eventually drew the line. Continuing solo in almost zero visibility, I found myself committed to a demanding obstacle course. First a rock crack, to be ascended only by bare hand jamming; next, a 'bad step' – that is, a leap across a seemingly bottomless cleft; and then a tower of boulders, to be attained only by a series of mantelshelf press-ups on rock covered in black lichen, which finally found me perched dizzily on the single summit cap stone. This picturesque summit of Gessi (15,418 feet) is the lesser of the two, Iolanda, at the south end of the ridge, being 50 feet higher. I had deliberately chosen the northern peak in the faint hope of obtaining the view that I felt was mine by right – of Lake Albert, 60 miles to the north, from which I had first seen the mountain over forty years earlier. Alas, the unrelenting cloud robbed me of this: I knew only too well that I was at the heart of that 'dense grey-blue mass' that had deluded the early travellers, but I took some consolation in feeling that I had kept faith with Romolo Gessi.

Back at Bigo, Moses, Peter and I, with thirteen carriers, set out on our final journey to the north, to complete our traverse of the range from end to end. We had no one in our party who claimed familiarity with our proposed route to Bwamba, but old Kajangwa had hunted over some of the region we would pass through.

In fine weather we made our way northeastwards up the broad and pleasant Bukurungu valley with its exquisite heath-moss woods and, higher up, forests of lobelias. But the pass itself is a wearisomely extensive area of bog tussock. The Portal range, showing us now for the first time its western aspect, was on our right hand. Passing the shallow East Bukurungu lake, we thankfully climbed a few hundred feet above the bog into a large groundsel-filled re-entrant in the range, where Kajangwa led us to a good rock shelter at 12,900 feet, with just sufficient room, 100 above it, to squeeze my tent on to mossy rocks.

Early in the morning Peter, Moses and I started a spirited attack on the Portal ridge. It turned out to be an even worse version of Gessi. We entered a gully crammed with all the horrors that Rwenzori can devise, including a wet and slimy rock chimney, the upper part of which we overcame by inelegant floundering in which all the conventions of climbing were flouted. It did, however, lead us to the final crest which, even in the prevailing mist, turned out to be a place of great charm. In between its giant rock *gendarme* features there was space to walk about amongst pretty ponds and moss gardens pleasingly evocative of the Hebrides. Indeed, the Portal range could perhaps be described as an enlarged and vegetated version of the Cuillins of Skye.

We had no difficulty in attaining Middle Portal peak, which at 14,337 feet was all our sense of duty called for, and after unnecessarily effusive handshakes and a few roasted groundnuts, set about the descent in dense mist. I was tired and when we entered the difficult moss chimney section, my concentration was

weakening. A slimy moss foothold lacked friction and I found myself precipitated downwards in a helpless free fall. By good fortune, I jammed on some rocks about 12 feet below. It doesn't sound very far, but the acceleration and impact were quite shocking and left me temporarily dazed. The main crunch between me and the rock had been taken by my Camera Care pouches and the cameras and lenses they contained, and when Peter had extricated me and I had found that no bones were broken, I anxiously checked my equipment and found all well – a tribute to Camera Care and Nikon!

Rather stiffly I eventually reached camp safely and it was only the next morning that I realised that I had been left with a strained back. The day's march was purgatory – I think the worst I have ever experienced. Quite apart from my painful back, it presented what is possibly the longest stretch of carex bog in the range, along the course of the Kisindika stream, where the depth of icy black water in the swamp forced us into prolonged stretches of tussock-jumping. These tussocks are particularly well developed and wobbly – no fiend in hell could have devised a more agonising exercise for a strained back. To complete this nadir of Rwenzori travel, it settled in to rain heavily and this persisted all day long, a day in which we covered 2½ map miles in six hours.

By mid-afternoon we were at the confluence of the Kisindika with the Lamia, one of the major rivers of the range, which has its sources on the eastern slopes of Mount Gessi. Soon after, to my infinite relief, we left the bog country behind and found ourselves moving through heath woods. Within half an hour I heard whistles of amazement ahead, and found my gang excitedly gathered around a little house – an elongated beehive house made of carex reed, the thatch reaching to the ground. It was still tippling down with rain, and we all crammed into the house, sitting on each other's laps, while a lichen fire was kindled that nearly choked us with its acrid yellow fumes. 'A smuggler's house,' said Kajangwa. 'This is the

Opposite. *Bamboo and mimulopsis tangle in the Nyamwamba valley. While our path-finders are out ahead, the rest of us gratefully take a rest.*

way the smugglers come from Zaire.' (It was an accepted fiction in conversation that smugglers always came from Zaire, not Uganda!)

Whatever its origin, it had evidently been there some time, because galium vines and alchemilla had grown over the thatch: it was like something from Peter Pan or Hansel and Gretel. We sat it out, as the rain pelted down and smoke and steam oozed up from the damp thatch. But before sunset the rain stopped and we emerged, to find that the smugglers had chosen a nice level patch, suitable for our tents, while nearby was a crystal-clear streamlet. When the brilliant stars filled the now cloudless sky, this turned into one of our most memorable camps.

With the bemusing fickleness that is so characteristic of these maddening mountains, this proved to be a turning point. From now on the sun shone and our journey became a celebration of all that is beautiful. The Lamia proved delightful, a bubbling, whisky-coloured stream growing stronger with every step we took, its water meadows carpeted with flowers. Above all, the giant heathers were massively in flower. At over 11,000 feet insect life was abundant, and I found clusters of fritillary and small blue butterflies feeding on patches of blue flowering thistles.

Our next camp was on the edge of the hagenia belt, and the following morning, within half an hour of setting off, the men discovered a bees' nest in a giant old hollow hagenia tree. Everything was dropped while a densely smoking lichen fire was lighted in the hollow trunk, and the heroic Muhindu, stripping off his shirt, climbed up with his *panga* and started urgently chopping a passage in to the nest. Oblivious of the bees, which were swarming about us in their myriads, the others stood below offering advice. Within an hour, improvised containers were being passed up for

Muhindu to fill with masses of the comb. We could not resist gorging ourselves there and then: after weeks of unvarying safari diet, this high-altitude heather honey was a celebration.

We followed the Lamia to the point where it enters Zairean territory and then reluctantly left it, climbing despairingly steeply through freely flowering mimulopsis on to the great ridge covered with bamboo forest that leads towards Ugandan Bwamba. We now had a recognisable hunters' track, and in several small clearings we found their shrines. These are miniature bee-hive houses, 1 or 2 feet high, made of bamboo sheath thatch, in which offerings such as an egg or banana are placed, to solicit the goodwill of the spirits of the animals. Sometimes above them there are stakes 3 or 4 feet tall with a vine stretched between them, from which little offerings are suspended. These invocations to the Spirit of the Mountains, Kitasamba, and the Guardian of the Animals, Kalisia, are a guarantee of good hunting. I do not know how Kajangwa met his religious obligations, but he was certainly successful. He scarcely ever failed to provide us with a morsel of meat to sweeten our diet, often catching his first hyrax within an hour of pitching camp.

At last we broke out of the bamboo belt and, our every step growing stronger with the

Below. *Hunters' shrines made of bamboo sheath, in which small offerings of food are placed to solicit the goodwill of* Kalisia.

increased oxygen and warmth of lower altitude, our last day's march – an unbelievable 6 miles in 9 hours – brought us to a pioneer clearing in the forest, and not far below this a little Konjo homestead at 6,600 feet. Although our journey was far from over – we still had to descend into Bwamba and the Semliki valley, and then cross the northern spur of Rwenzori to make our way home to Ibanda – none the less our traverse of the range was complete.

This homestead was on a ridge, surrounded by banana gardens. Towards sunset I walked a little way round the side of the ridge and came to a place from whence a superb view opened before me. At my feet, the forest-bare cultivated wrinkles of the foothills dropped steeply down to low-lying Bwamba – a patchwork of cultivation 3,000 feet below. This gave way in the middle distance to a measureless zone of rainforest canopy stretching between the western and northern horizons – the easternmost reach of the Ituri pygmy forest of the Zairean basin, through which Stanley's column had struggled a hundred years earlier. Through this meandered tortuously the great Semliki river, recognisable by the darker colour of the riverine forest galleries. As I watched, the nearly horizontal rays of the setting sun flooded this section of the Rift valley with golden light, throwing into relief, on my right hand, the green hills of Rwenzori's last spurs, and on the left, the blue hills of the Zairean escarpments. In between, a little east of north, the horizon became a sheet of shining silver. This was the antithesis of the 1943 experience with which I started my story – the inland sea, Lake Albert, seen from the mountains.

After dark I returned to my viewpoint, from which the vault of heaven now presented a crowded canopy of stars of a brilliance such as is not to be seen from our northern industrial lands. Below the lowest stars in the western sky a belt of blackness showed where the forest lay, while at my feet, far below, were scattered red pinpoints of light – the home fires of the

Baamba, eloquent symbols of man's insignificance in the scheme of things. Unlike the steady brilliance of the stars, each of these tiny fires blinked and flared uncertainly, some vanishing not to return. There was a pathos about it, a sense of the frail hold that man has on the African earth.

I stirred early next morning and was at my vantage point before the sun had touched the western escarpments. The air was peculiarly clear and Lake Albert lay like a sheet of frosted glass across the northern skyline. But what struck me most strongly was the behaviour of the atmosphere. Over the Rwenzori spurs a delicately pencilled curve of semi-transparent vapour hung – no more than a faint diaphanous whisp; and the same was more distantly layered over the western hills, while the course of the Semliki river was marked by a similar wafer-thin haze.

The impression was of the extreme thinness of the earth's atmosphere – a mere whisp that lies between us and the blazing, sterilising sun. Yet within that whisp all living processes are contained – the whisp is our biosphere.

Diagrams illustrating the earth's atmosphere often imply that it is synonymous with the biosphere and show the earth comfortably wrapped up in a coat that, to scale, would be hundreds of miles thick. How thick is this coat really? The biosphere, in which all life occurs, is as deep as it looks. In your garden, it is a few feet deep; in a tropical forest, a few scores of feet; in the oceans, a few hundred feet. To get a true perspective of the ultra-delicacy of our biosphere and atmosphere, consider the following analogy. Take a school globe of about 2 feet diameter: wrap it tightly in the thinnest cling-film such as is used for wrapping food, and assume that this film is one-twentieth of a millimetre thick. Such a film would represent a

Opposite. Scadoxus. *This amaryllid (often called* Choananthus *in early botanists' reports) grows in generous patches in the upper belt of the rain forest at altitudes of 8-9,000 feet.*

Scadoxus cyrtanthiflorus

⌐1cm⌐

× 1/3

biosystem about half a mile thick! In short, our practical biosystem is almost unimaginably thin: it is no more than an interface between solar bombardment and a *two*-dimensional world. All human hopes and endeavours are constrained in a film proportionately much thinner than the green algal dusting on a tree's trunk. This is the measure of the desperate frailty of our only home, planet earth; and of the fatal complacency that on the one hand we can abuse that biosystem as we please, while on the other, we allow the unrestricted increase of the human race.

Within half an hour of sunrise, my view was fading in the haze of another tropical day. I returned to my tent, took out my note-book and started writing. From those notes, written while the spirit of the hills was still about me, have grown the concepts that form the last chapter of this book, which spells out my views on the last chance we have to save Africa, the Nile and the Mountains of the Moon.

Below. *Descending the north-western flanks of Rwenzori above Bwamba, we came across this pioneer forest clearing at about 7,500 feet. The 'owner' was hard at work, and told me he would plant cabbages and beans as his first crop. This is an example of the piecemeal way in which Rwenzori is losing its forests. The trash will be burnt and the high-humus but fragile and thin covering of the forest earth will rapidly be lost by erosion, leaving an impoverished soil that is useless as a water catchment.*

LAST CHANCE IN AFRICA

The ineluctable laws of population dynamics will guarantee the total denudation of Africa's Mountains of the Moon up to the highest altitude at which human ingenuity can grow food and extract firewood. The response of Western minds to such a prospect is to think in terms of declaring a national park or World Heritage Site. As we shall see, this is what I propose, but the problems facing Rwenzori and other special areas of the continent will only be solved for the long term by addressing the basic problems of Africa as a whole. A form of holistic approach is called for, and I make no apology for now turning from the smaller problem to the greater.

Most Western endeavour in tropical Africa has failed with more or less damaging effects on the culture and morale of the indigenous society and its environment. It has been enormously wasteful in terms of cash, resources and well-intentioned enterprise. Sir Michael Wood, founder of the East African Flying Doctor Service, with the experience of a lifetime devoted to African welfare, has recently written, 'The first two decades of aid . . . have been largely a failure. The projects . . . are no longer operative a few years later. Huge sums for relief . . . have not been used for the ends for which they were given . . . Technology was introduced where there was no hope of maintenance or practical application . . .'

Only in very recent years have the increasingly disillusioned Western aid analysts started to speak out frankly. Essays such as *Uganda Now* and *Tanzania: crisis and struggle for survival* are necessary reading for those who wish to understand the nature of the failure of the Western development concept. Modern Africa is a white-elephant graveyard of the failed and inappropriate. As you walk through this wasted landscape, almost the only things you will see that really work and contribute directly to the daily lives of the African people are the original systems that we Europeans found when we first came to these countries in the last century and which were still largely intact when we left a quarter of a century ago. These are the systems which are so finely tuned to every varying corner of the countryside, that have allowed human life to succeed over past centuries in the most unpropitious environments, while yet sustaining a balance with that environment and not diminishing it. Even as late as the 1940s, when I first went to Africa, the destabilising process had scarcely started. Nearly all the countries that are now at risk were then self-sufficient in staple foods.

Africa's so-called primitive husbandman is far more closely adapted to his piece of land than his counterpart modern farmer in Europe or America. His intuitive local knowledge is one of the few things of value that he has, and he stands to lose disastrously by its disruption. The Western agricultural techniques which have evolved in our benign rain-gifted climate have failed almost universally in Africa. Here, far from developing the cultivable loam that is the basis of Western agriculture, such methods lead to impoverishment and erosion.

Before the introduction of European thinking, Africa had developed its own approach to its special problems – long-interval rotation. Under this, land was used for a few years for small-patch, mixed planting of crops such as millet, beans and squashes, all side by side in the same little garden, thereby reducing pest and weed problems, protecting the soil from rain impact and improving soil stability. Such gardens were abandoned when yields began to wane and left for several decades as fallow. In this way the population, obviously smaller at that time, struck a balance with its ecosystem and preserved the precious vegetational mantle of the earth. Since then almost every consequence of European contact has been to destabilise that vital balance.

We vets and agriculturalists have done more harm than good by our attempts to introduce inappropriate crop and livestock systems. We have triggered off uncontrolled increase in the human and livestock populations while scarcely increasing the *reliably sustainable* supply of food. Our medical colleagues have equalled us. They have been assiduous in controlling diseases without considering how the resulting escalating populations are to sustain a balance with their environment. Life-saving has been seen as an end in itself: but it is not an end, it is only a beginning.

Dr W. E. Ormerod, lately of the London School of Tropical Medicine, has courageously spelt out this dilemma. 'I believe that . . . mass disease eradication campaigns will make the situation much worse, and greatly increase the mortality from starvation and other diseases . . . the consequences will involve destabilising the balance African farmers have obtained with their environment . . . If we enhance productivity in a way which fails to maintain stability, the environment of the African husbandman will rapidly be degraded and his interests will have been ill served.'

The ten years that have elapsed since those words were written have amply fulfilled the prophecy. We Europeans only had the right to reduce mortality if at the same time we ensured the well-being of the larger populations that our actions created. We have failed to do this and having abrogated this responsibility we carry a grave burden of blame for the desperate state in which Africa now finds itself.

Dr Michael Wood asks the same question: 'Are we keeping people alive today for them to die next year?' He goes on, 'These are harsh questions, but they are being asked. African governments need to give the answers.'

A quarter of a century has been lost, but there are signs that lessons are at last being learnt, and the one-time self-assurance of the aid industry is nowadays more muted. But this new realism so far extends only to the more obvious aspects of the crisis. There is a more pernicious misdirection which is never mentioned. This is the damage which contact with Western culture continues to do to the minds of the indigenous people. Whether peasant schoolchild or urbane graduate, Africans are insidiously fed the impression that homely methods are inferior and that advance must be by Western wizardry. Language itself prevents thinking. Transient expatriate aid workers, unlike the career men of the colonial era, seldom speak a vernacular language, so the exchange of ideas is restricted to those of the local people who speak English. These constitute that élite class of Africans which, in the process of learning English, has also been subconsciously indoctrinated with the Western way of thinking. A comforting closed circle pertains in which the European expresses his Western-oriented concepts which have been derived from his temperate climate experience, to African minds already primed by the very language itself to accept them. As with George Orwell's 'newspeak', you can only think with the words and meanings you have been provided with. In the process of his accepting them, the African recipient's ready response in familiar language warms the European speaker with the reassurance that he must be right. Meanwhile, during the decision-making, the person most deeply involved, the inarticulate non-English-speaking peasant on whose response success or failure will depend, is ignored or denigrated, even though he may have, from his ancient roots, vital wisdom to contribute.

It should be realised that we Europeans simply do not have ready solutions for problems that we ourselves have largely created. This is a bitter pill to swallow and it runs uncomfortably counter to the warm emotionalism, learnt from the Victorians, that calls on us to feed the hungry, treat the sick and instruct the so-called ignorant. Such motivation, if we would but admit it, may be more concerned with assuaging the feelings of guilt of a people who have created for themselves a starvation-free and affluent society, rather than with trying to understand the complex nature of other, and by inference inferior, societies.

It is a basic rule of ecology that a species will expand up to the point at which limitations on its nutrient supplies impose restraints. The West's impact on Africa is largely geared to increasing food production and reducing disease. The result has been to take the brakes off a potentially fecund population, so that when a series of years of adequate rains raises the ceiling of nutrition, there are simply more people little better off then before. When inevitably the rains fail, this increased population is caught out of balance with its food resources, with results that are all too familiar on our television screens. Already half the population of most sub-Saharan countries is under the age of 16. This means that, in the immediately coming years, an enormously enhanced proportion of the population will be bearing children compared with stable populations elsewhere. The laws of population dynamics dictate that such a ratio, in the absence of a sensible compensatory controlling mechanism, comprises a biological time-bomb fused to cause irremediable damage by the end of the century.

The way in which Western culture contact, with all its good intentions, is essentially causing the destruction of what it aspires to help, is complex. Apart from the manipulation of food supply and disease control, there are other influences. Prior to the arrival of Europeans, there were important aspects of African life that tended to control population growth and so sustain the ecological balance. Such factors were bound up in a code of social behaviour which, before the dismantling of tribal law and custom, was rigorously followed. It is important to look at some of these factors and consider the effect they may have had on the birth-rate.

First, in their early teens, groups of pubescent girls were instructed by elder women in sexual know-how, the arts and permissions of safe sex play, and the essentials of avoiding pregnancy. A strong social stigma was carried by pre-marital pregnancy. When a young girl became the subject of an offer of marriage, the contract involved the payment of a dowry or so-called bride price by the husband-to-be's family. Essentially this consisted of livestock – perhaps ten or fifteen cattle – finding such wealth might take a long time, thereby delaying the commencement of a woman's child-bearing life.

When the wife gave birth, there was a proscription on the resumption of sexual intercourse so long as the baby was being breast fed. In pre-European days breast feeding was universal, over periods of two or three years, and this system effectively spaced out a woman's child-bearing, benefiting the health of both mother and baby. This reinforced nature's own form of contraception, the physiological mechanism whereby a woman's ovulation remains inhibited by suckling. The consequence must have been a significant check on the overall birth-rate of such societies.

Europeans introduced the now widespread practice of bottle feeding and one of the saddest sights of modern Africa is the well-meaning, misinstructed mother feeding her baby with a beer bottle and teat made of a maize cob. The mixture is likely to be misappropriated aid agency dried milk sold corruptly on the black market, made up with unsterilised dirty water, and quite unsuitable as a baby diet. The result is malnutrition, diarrhoea – and the early fresh pregnancy of the mother.

What we have witnessed is the abandonment of the old constraints on population, *without their replacement by modern alternatives.* Those of us who feel a sense of responsibility about this crisis must accept that, whatever its origins, population escalation is now the over-riding factor, and if we fail to address it squarely, we will fail in everything. Yet there are those who speak as if there is something improper in advising families or nations to constrain their procreation, although they themselves speak from the secure base of a

society that has stabilised its own population.

A false analogy is sometimes drawn from our European experience of population expansion during the eighteenth and nineteenth centuries, but this is not relevant to the present African scene. We did not experience an explo-sion, but rather a comparatively modest six- or seven-fold increase over something like two hundred years. Compare this, for example, with Kenya's ten-fold increase over sixty years, to become twenty-fold after only another twen-ty years. The circumstances were of course

Opposite. *In Rwanda, on the steep slopes of Mount Sabinyo, the montane forest, once range-land of the mountain gorillas, has been excised from the national park and totally destroyed to permit cultivation. The picture shows denudation right up to the line of the bamboo forest at about 8000 feet. In the foreground, pyrethrum fields abutt a home-stead of the Banyarwanda. In the mid-ground they will attempt to grow food crops, with inevitable consequential soil erosion. The pale blue-leaved saplings and tall straight darker trees are introduced species of eucalyptus, planted for firewood and building poles.*

completely different. The Western increase took place in a world devoid of the products that only we could invent and make; it was facilitated by our mastery of the sea and the development of North America as an inexhaustible bread basket, and sustained into the twentieth century by our own agricultural and technological revolution. By comparison Africa produces little that the rest of the world needs.

The view has been widely canvassed that Africans have large families as an insurance policy against high infant mortality and old age. This needs to be examined, because the extent to which it is true has probably varied over past years. Before the traditional clan structure broke down under Western contact, it provided a firm social security system and the concept probably lacked validity. However, the Pax Britannica led to a widespread dispersal of homesteads throughout the empty and now peaceful countryside and as long as land remained plentiful, large families tended to replace the clan system as a provider of security. But in present times, with land shortage under excess population and extensive urban migration, while large families continue to be the vogue it is questionable whether social security is the main reason, rather than the underlying driving force of male machismo. If the *women* were fully informed and able to exert free will, I have no doubt that they would opt for smaller families. Chinks of light are only just beginning to be cast on this aspect of African life, but heart cries such as Awa Thiam's *Black Sisters, Speak Out*, are I believe indicators of a tidal wave of African feminism that must inevitably come.

Another assertion, that the birth-rate will fall

as living standards rise – even if it was theoretically true – is seen to be pointless in conditions in which it is the high birth-rate that is preventing such a rise and is indeed causing standards to fall still further. I say theoretically because, while this is a correlation that has been observed in the West, it is not necessarily a case of cause and effect. The assumption derives from Western reasons for having children. African ambitions are different to the extent that in this male-dominated society a man is judged by the number of children he has; his reaction to increasing prosperity, far from leading him to limit his family, is more likely to lead him to increase it.

People sometimes speak of the immense size of many of these countries and their relatively light *average* population densities compared with Europe, and then make specious predictions that the continent could support enormously greater populations than it presently does. The argument is fallacious for two reasons. First, the maps are delusory: much of the 'unused' space is in fact uncultivable, rain-deficient near-desert. Second, even if at some future stage, by presently unknown techniques, Africa could be made to support a much-increased population, would it not be irresponsible simply to let the population rip on the basis of such guesswork? It is our duty to try to reserve for future generations the choice between more people and more per capita well-being.

Our experience in the West is thus seen to be a misleading indicator for the options for Africa. What is self-evident is that our Western affluence and well-being could not be sustained without our controlled birth-rate. We have signally failed to make it clear to the Third World that, whatever may have happened in our past, population stability is the keystone of our present enviable (by their standards) state. To the extent to which we have failed to get this message across, we have been guilty of deception. We have only to look at the graph on page 163: this is incontrovertible; whatever the causes, the rights and wrongs, these countries and others like them are heading for biological cataclysm.

In order to put the information presented in the graph into perspective, I have included a Western country that at the start of the period shown had a comparable population. In 1925 Norway had almost exactly the same population as Kenya. Sixty years later, her population has scarcely increased by half; it is now virtually stable and her people enjoy a degree of social well-being which is hardly surpassed elsewhere in the world. In the same period, Kenya's population has multiplied ten times, and in spite of all the commendable efforts of the Kenyans, no substantial impact has been made on the widespread poverty, while the country has suffered enormous environmental deterioration.

Because it is difficult, and for some people apparently embarrassing, we have lacked the moral courage to set out the correct agenda. At the top of the list should be population containment, leading to environmental stabilisation, leading to the development of a sustainable agriculture. In the past, Western agencies and media have conspired together in a fog of euphemisms. The knee-jerk response to Africa's plight – whether from a TV programme, parliament or pulpit – is to demand expansion of financial aid, and I dare say it would be generally approved if I were to end my requiem for Africa's Mountains of the Moon with the customary appeal for more development funds.

This is not my intention: far from expanding financial aid I believe we should be restricting it to those quite narrow fields where we can say with confidence that it will be beneficial in the long term. New aid schemes should be strictly assessed on the criteria of how they meet the related problems of population increase and environmental deterioration. Research and educational programmes related to environmental management and family planning should rank far above conventional subjects such as food production and medicine.

Although it is indisputably the most important question in Africa today, the problem of how to tackle over-population has been almost totally ignored, whether by the indigenous governments or the donor governments and agencies. I believe we can no longer delay attempting to devise radical policies for population stabilisation – policies that are likely to have an immediate effect on the rocketing growth curves of the graph shown here.

Here, as so often, we find ourselves blinkered by our Western experience. From the invention of the condom to the most recent advances in endocrinology and immunology, the tactics of the Western sexual revolution have seemed so obvious to us that the question of a necessary background strategy has scarcely arisen. However, attempts to transplant these tactics to Africa have so far largely failed. The reason for this is the widely different ethos of African society. In a situation where the male's machismo demands children, family restriction advice addressed to men is unlikely to be heeded; and if addressed to women, is only likely to be effective if it is in a form that is going to put them in a stronger position in the face of male dominance. The problem thus becomes a question of what can be done for African women, not just to provide them with a contraceptive option, but to provide an incentive to take up this option that is stronger than the strong male pressure to ignore it.

The aims should be to delay the age of marriage; to delay pregnancy, whether within marriage or without; to extend the intervals between pregnancies; and to offer women an alternative to the present unquestioned treatment of them as a reproductive machine. At present there is only one time in an African woman's life when she is in a position to call the shots: this is the pre-marital period, when the dowry is under negotiation. I suggest that she should be offered some form of personal wealth that could on the one hand give her negotiating power in the choice of a husband or, on the other, finance the option of delaying or refusing marriage. The extent of such wealth and the manner of its provision would be dictated by what was found to be necessary to outweigh the existing social pressure for her to marry early.

I propose that the State, riding in a sense on the back of the traditional dowry system, should introduce a 'State Dowry'. Under this, every woman on reaching the age of, say, 16 without becoming pregnant, would become eligible for a regular cash allowance. The payments would cease as soon as it was established that she was pregnant, but would resume again after a further period of non-pregnancy. There would thus be a strong incentive for her and her parents to delay marriage, and after marriage, for her and her husband to delay conception.

Would such a system be unacceptably expensive? One way of answering this is to say that it would effectually cost nothing. All that we would be doing would be to alter the direction of the flow of money. At present, money flows from the West to governments or agencies, and thence through the hands of ministries and departments, to contractors and managers who are all too often incompetent or corrupt, with little benefit to rural communities. This procedure, whereby multi-billion dollar sums are poured away, must be regarded as unacceptably expensive, wasteful and demoralising. Under the State Dowry System, the same sort of volume of cash would go directly to peasant families through post office accounts and could scarcely be diverted *en route*. For the first time, the African agriculturalist (and women are the primary agriculturalists) would be free to decide for herself how she wished to be 'developed'. Whether more or less wisely saved or spent (and my belief is that rural women would spend it wisely, provided they could keep it in their own hands), the money would find its way into uplifting the village economy, which is something most aid schemes signally fail to do.

Dr Raymond Crotty, of the School of Systems and Data Studies at Trinity College, Dublin, has called my attention to the similar proposal in his book *Ireland in Crisis*. His computations show that, on the basis of the current annual Western aid to the Third World, adequate funding would be available to halve the present birth-rate. This would, for example, meet the recent plea of President Moi of Kenya, to reduce the average Kenyan family from eight to four children.

While in the early days of such a system women might find it difficult to ensure absolute control over dowry payments, the very existence of the scheme could act as a stimulus to the advance of feminine power and the development of women's organisations. These would eventually lead to recipients achieving a large measure of control over funds paid in their name. A possible outcome of this could be a moderation of the present oppression of women and a hardening of their attitude to such institutions as child marriage, polygamy and female circumcision.

Such an idea would obviously take off slowly and cash requirements in the early days would be small. The concept could of course be pilot-tested on a scale as small as desired in order to find out the best administrative system. It is easy to point out difficulties: the solutions must be found by trial and error. It will be appreciated that we would in effect be creating a child allowance system without the incentive to have children that such a system entails. The funds paid to discourage women from starting a pregnancy would in fact enhance the welfare of any existing child, or a child to come in the future. The chances of survival of such well-spaced and well-provisioned children would be much greater than is commonly the case at present. One may note that while all societies accept the concept of paying people to do things, we in the West increasingly pay people *not* to do things. The principle has been widely used for controlling agricultural production and for encouraging conservation policies. In the final analysis it will be seen that we would simply be reinforc-

ing every woman's inalienable right to say no.

The actual logistics of providing the necessary family planning services are sometimes spoken of as a stumbling block, but I do not share this view. We should get away from the idea that sophisticated clinics with highly trained staff are necessary. Village shops, free dispensing machines and barefoot advisors should provide a service at the very simplest level. In the early days, we should not feel behoven to provide the near 100 per cent guarantee against conception that is the norm in the West. For reasons that I have made clear, in the early stages the widespread adoption of the male condom would be unlikely, but recent prototype development of a female condom points the way to future possibilities.

Any discussion of future population trends in Africa would be unrealistic if it did not take into consideration the dark shadow of AIDS that stretches across the continent as a heterosexually transmitted disease: my proposals for the dowry system should be looked at against this background. The future effect of AIDS is difficult to assess, but a recent attempt by researchers at Imperial College, London, and Princeton University, USA, concludes that it is unlikely to make any major difference to current African population forecasts for several decades to come. There is to my mind one hopeful chink of light. The only measures that can be anticipated to control the AIDS pandemic are largely the same as those that will be necessary to bring about population stabilisation. AIDS in the West will be controlled in the end by the polarisation of society according to behaviour patterns. On the one hand a sub-culture will emerge which, because it adopts a preventive approach to sex, will be virtually 'immune' to this avoidable condition. At the other pole, the disease will run its tragic course.

It should be noted that eligibility for the State Dowry would be universal, for town dwellers as well as country folk, and it could be expected that it would become a major factor in promoting safe sex and the use of the male and female condom as well as reducing prostitution and promiscuity. African urban prostitution is a response to urban poverty and male mobility and the widespread abandonment of wives and partners that has become a feature of the unstable new urban societies. By providing such women with an alternative source of finance, I believe that the forcing of them into prostitution would be reduced.

I am of course aware that many people will find the ideas expressed in this chapter difficult to accept. Such people, understandably, will mainly be those who have no first-hand experience of grass-roots Africa and have not had the opportunity of coming face to face with the

escalating environmental destruction and human misery that is being caused by the population explosion. I respect such people's sincere beliefs but at the same time, I suggest that they have a moral duty to examine the facts – indeed, perhaps all they need to do is to look at the graph opposite – and then suggest a better practical way forward that will effectually alter these population projections before the continent has passed the environmental point of no return.

In the past, both the environmental and the aid lobbies have remained largely silent on the critical relationship between development and the environment on the one hand, and population on the other. But this is changing: since I

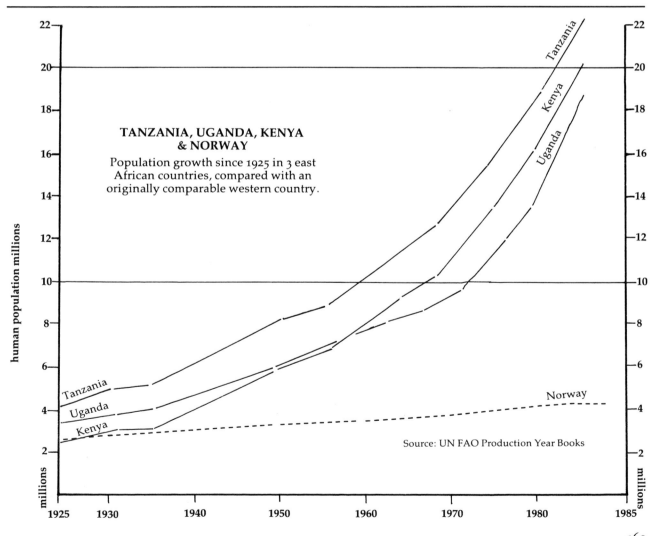

TANZANIA, UGANDA, KENYA & NORWAY

Population growth since 1925 in 3 east African countries, compared with an originally comparable western country.

Source: UN FAO Production Year Books

human population millions

drafted this chapter I have been enormously encouraged by the authoritative voice of Edward Goldsmith, editor of the *Ecologist* magazine and one of our leading environmental philosophers. His bold and courageous leading article in a recent edition of this magazine, in which he addresses the taboos that have for so long fudged this subject, perhaps ushers in a new era of rational thinking. Recent forthright statements grasping the population question by such people as Sir Peter Scott, Dr Jacques Cousteau, Dr David Bellamy and the Duke of Edinburgh perhaps indicate a sea change in the current of world opinion.

I have come a long way from my declared purpose of protecting the Mountains of the Moon and the sources of the Nile by making Ugandan Rwenzori a national park and World Heritage Site. The path is indeed long, but it is the only one worth following if we are seriously going to try to conserve not just a few attractive habitats but the whole complex range of Africa's biosystems. All our good intentions will be fruitless if we cannot first control ourselves. All wildlife and environmental conservation will in the end be pointless and will fail disastrously, if mankind is to be allowed uncontrollably to pre-empt the resources of the biosphere. The transparently thin film over the surface of our planet, our only living space, is indivisible.

Below. Under Belgian tutelage, deforested mountain slopes such as this one in Rwanda were subjected to contour ridging to try to limit erosion. Abandoned at first after independence, the agricultural department is now trying to re-instate the system. But the strips are not levelled to form true terracing, such as one sees in eastern Asia, and serious soil loss still occurs in the rainy season.

Patching operations here and there can only be temporary expedients, while we grasp the heart of the matter – that we share a two-dimensional world of limited and diminishing resources.

If, as I have suggested, the long-term protection of the jewels of Africa can only be assured by policies that will protect Africa herself, none the less in the interim we have a duty to do what we can at the immediate and local level. All the wildlife parks and reserves of Africa are going to have to struggle to survive under the population tidal wave of the next few decades. I have no doubt that what we have seen happen in Rwanda will occur increasingly elsewhere. The existing national parks of East Africa were nearly all founded in the days before the African crisis was recognised, on the principle of total human exclusion that saw poaching, rather than population, as the threat.

As a consequence, they have got the worst of both worlds. Total exclusion of human activity leads, at the very least, to an indifference amongst the surrounding population, and, as land shortage becomes more acute, to bitterness and hostility. In such circumstances it is not possible to inculcate the local climate of support that is the only permanent means of controlling poaching. One can predict that the wildlife parks of Africa will only survive to the extent that the local people want them to

Below. *No effective way of re-establishing 'natural' African montane forest has ever been developed. This picture of natural forest remnants on the Nile-Zaire (Congo) divide at about 7000 feet was taken in Burundi. The American funded re-afforestation project is experimentally attempting to 'knit' the remnants together with non-indigenous conifers, in the hope that eventually the natural species will re-colonise these areas.*

survive, and this may mean a progressive degree of integration of human activity with park function, including the controlled harvesting of the wild game.

This prediction should be borne in mind in considering the rather different question of the mountain parks. The existing parks on Mount Kenya, Kilimanjaro and the Nyandarua (Aberdares) are misleading to the extent that the main protected areas lie above the tree line, leaving the zone that is most at risk – the lower mountain forest belt – at the mercy of politically susceptible agricultural and forestry policies. Irrespective of such precedents, I believe that, on the one hand, if we do not conserve the lower deciduous forest belt of Rwenzori we shall have failed tragically, and on the other, any proposal to exclude the Konjo totally from what is essentially their own tribal domain will be equally doomed to failure. A new form of national park should be envisaged, a 'forest park', in which much of the forest zone of Rwenzori would be managed by the Konjo themselves, under guidance in the early years, as a sustainable and renewable resource.

This is the basis of Dr Peter Howard's submission to the Government of Uganda, on behalf of the World Wide Fund for Nature and the New York Zoological Society. Understandably the indigenous Ugandans who are striving to restore normality to their so gravely disrupted administration and economy cannot be expected to give high priority to a concept that must seem far removed from their day-to-day pressures. None the less, the advent of President Museveni and his new administration must be welcomed as the most hopeful development of recent decades, and it is surely up to those of us in the outside world who see the need and have the resources, to help Uganda to preserve her priceless treasure. I believe it is now up to the EEC environmental programme and USAID to come forward and underpin the proposed new park. At the same time, the International Union for the Conservation of Nature should declare Uganda's Rwen-

zori a World Heritage Site, continuous with that already covering the Zairean sector. Only by doing this will Europe be keeping faith with the Ancient Greeks.

Perhaps we should not be talking about a 'national park'. We are really talking about the Konjo domain – a domain to which, in spite of historical injustice, they have shown themselves to be so true. Their heritage and culture are inescapably bound up in these incredible hills that have for so long mysteriously gripped the imagination of the rest of the world. Could it be that the recognition by Uganda and the world of a unique dynamic *Konjo Heritage Domain* would be accepted by the Rwenzururu irreconcilables as sufficient recognition, thus finally fulfilling their understandable aspirations? And should not we in Britain, who must bear historical responsibility for the initial injustice of a hundred years ago, offer our goodwill and assistance in bringing this about? Perhaps the solutions to both the political and environmental crises of these ultimate sources of the Nile lie in one single imaginative act of faith.

Opposite. Thunbergianthus ruwenzoriensis. *Christabel King found this handsome climber beside our path below the forest as we approached the Nyabitaba ridge at about 6000 feet.*

Overleaf. *A minor valley in south-eastern Rwenzori, with Konjo homesteads and gardens. The valley floor is at 4,000 feet, the highest ridges 7,500 feet. Virtually the whole of this belt has been denuded of forest during my working life. To the informed eye, the intensive unterraced cultivation of the steep thinly soiled slopes and the numerous signs of erosion spell the advancing destruction of this once beautiful environment and its ruination as a water catchment.*

Thunbergianthus ruwenzoriensis

1 cm

GLOSSARY

Plants, mammals and birds mentioned in the text.

Plants

Alchemilla argyrophylla (Rosaceae): ground cover
A. triphylla (ditto)
A. johnstonii (ditto)
A. microbetula (ditto)
A. subnivalis (ditto)
Anthriscus sylvestris (Umbelliferae): wild celery, gorilla food
Antitrichia curtipendula (Leucodontaceae): the most common tree moss
Arabis alpina (Cruciferae): rock cress
Arisaema ruwenzoricum (Araceae): snake's head arum
Arundinaria alpina (Gramineae): mountain bamboo
Begonia meyeri-johannis (Begoniaceae): mountain white begonia
Breutelia subgnaphalea (Bartramiaceae): common ground moss
Canarina eminii (Campanulaceae): bell flower, climbing epiphyte
Cardamine obliqua (Cruciferae): lady's smock
Carduus ruwenzoriensis (Compositae): thistle
Carex runssoroensis (Cyperaceae): large bog tussock
Cyathea deckenii (Cyatheaceae): tree fern
Cynorkis anacamptoides (Orchidaceae): terrestrial orchid
Disa stairsii (Orchidaceae): tall rose-coloured orchid
Dracaena afromontana (Agavaceae): palm-like tree
Ensete edule (Musaceae): wild banana, stemless plantain
Erica arborea (Ericaceae): tree heath
Eulophia streptopetala (Orchidaceae): terrestrial orchid
Festuca pilgeri (Gramineae): dry tussock grass
Galium simense (Rubiaceae): goose grass, cleavers, gorilla food
G. ruwenzoriense (ditto)
Gloriosa superba (Liliaceae): flame lily
Hagenia abyssinica (Rosaceae)
Haplocarpha ruepellii (Compositae)
Helichrysum guilelmii (Compositae): herbaceous everlasting
H. stuhlmannii (Compositae): everlasting shrub
Hibiscus diversifolius (Malvaceae): purple mallow
Hypericum lanceolatum (Hypericaceae): yellow St John's wort
H. keniense (ditto)
H. bequaertii (Hypericaceae): scarlet tree St John's wort
Impatiens runssorensis (Balsaminaceae): scarlet balsam
I. stuhlmannii (ditto) pink balsam
Lobelia bequaertii (Lobeliaceae): giant lobelia of bog areas
L. gibberoa (ditto) of lower montane belt
L. lanuriensis (ditto) of heath moss forest
L. wollastonii (ditto) of alpine zone
Lycopodium saururus (ditto) giant clubmoss
Marchantia polymorpha (Marchantiaceae): liverwort
Mimulopsis elliotii (Scrophulariaceae): dense shrub thicket

Peltigera canina (Peltigeraceae): large foliose lichen
Pennisetum purpureum (Gramineae): elephant grass
Peucedanum kerstenii (Umbelliferae): giant hog's fennel
Phillipia benguelensis (Ericaceae): giant heath
P. johnstonii (ditto)
P. trimera (ditto)
Phragmanthera usuiensis (Loranthaceae): red parasite of forest canopy
Poa ruwenzoriensis (Gramineae): dry tussock grass
P. schimperiana (ditto)
Podocarpus milanjianus (Podocarpaceae): Podo tree, evergreen conifer
Polypodium rigescens (Polypodiaceae): epiphytic fern
Polystachya kermesina (Orchidaceae): epiphytic orchid
P. nyanzensis (ditto)
Pteridium aquilinum (Dennstaediaceae): common bracken
Ranunculus oreophytus (Ranunculaceae): alpine buttercup
Rapanea melanophloeos (Myrsinaceae): previously *R. rhododendroides*
Rubus runssorensis (Rosaceae): Rwenzori blackberry
Rumex afromontanus (Polygonaceae): dock
Sagina afroalpina (Caryophyllaceae): semi-aquatic pearlwort
Satyrium robustum (Orchidaceae): terrestrial orchid
Scadoxus cyrtanthiflorus (Amaryllidaceae): scarlet amaryllid (previously *Choananthus*)
S. multiflorus (Amaryllidaceae): fire ball lily (previously *Haemanthus*)
Sedum ruwenzoriense (Crassulaceae): stonecrop
Senecio johnstonii subsp. *adnivalis* var. *adnivalis* (Compositae): common giant groundsel of Rwenzori
    var. *alticola*: Virunga
        var. *erici-rosenii*: lower altitude giant groundsel
            var. *friesiorum*: giant groundsel of western Rwenzori
Senecio × pirottae (Compositae): hybrid small groundsel
S. subsessilis (ditto) yellow flowered daisy
Sphagnum pappeanum (Sphagnaceae) aquatic moss
Streptocarpus ruwenzoriensis (Gesneriaceae): related to African violet
Symphonia globulifera (Guttiferae): scarlet flowered forest tree
Thunbergianthus ruwenzoriensis (Scrophulariaceae): climber of forest zone
Urtica massaica (Urticaceae): stinging nettle
Usnea species (Usneaceae): Spanish moss lichen
Viola eminii (Violaceae): Emin's violet

Mammals

Cephalophus nigrifrons: black fronted or red duiker
Ceropithecus l'hoesti: Hoest's monkey
C. mitis stuhlmannii: blue monkey
Colobus angolensis ruwenzorii: black and white Colobus monkey
Dendrohyrax arboreus ruwenzorii: hyrax
Gorilla gorilla beringei: mountain gorilla
    (ditto) *graueri*: eastern lowland gorilla
    (ditto) *gorilla*: western lowland gorilla

Hylochoerus meinertzhageni: giant Forest hog
Lemniscomys sp.: striped vole
Lophuromys sp.: brown vole
Loxodonta africana: elephant
Panthera pardus: leopard
Pan troglodytes: chimpanzee
Potamochoerus porcus: bush pig
Tragecephalus scriptus: bushbuck

Birds
Anas sparsa leucostigma: black white-spotted duck
Buteo oreophilus: mountain buzzard
Columba arquatrix arquatrix: olive pigeon
Corvultur albicollis: white-necked raven
Cynnyris regius: regal sunbird
Nectarinia johnstonii dartmouthi: scarlet-tufted malachite sunbird
Ruwenzorornis johnstonii johnstonnii: rwenzori turaco

BIBLIOGRAPHY

Aeschylus   525-456 BC; *Supplices* 559

Agnew, A.D.Q.   1974, *Upland Kenya Wild Flowers*; Oxford University Press

Akeley, Carl E.   1923, *In Brightest Africa*; Garden City, New York

Akeley, Mary L. Jobe   1931, *Carl Akeley's Africa*; Victor Gollancz

 – 1951, *Congo Eden*; Victor Gollancz

Alnaes, Kirsten   1969, *Songs of the Rwenzururu Rebellion* in *Tradition & Transition in East Africa* (Ed. P.H. Gulliver); Routledge & Kegan Paul

Amon Bazira   1982, *Rwenzururu: 20 years of bitterness*; Origin not stated, 14pp

Anderson, R.M., May, R.M. & McLean, A.R.   1988, *Possible demographic consequences of AIDS in developing countries*; *Nature*, 332, 228-234

Aristotle   384-322 BC; *Meteorologica*, 1, 13

Baker, Samuel W.   1866, *The Albert Nyanza; Great Basin of the Nile*; Macmillan

Baumann, Oscar   1894, *Durch Massailand zur Nilquelle: Reisen und Forschungen der Massai-Expedition des deutschen Antisklaverei-Komite in den Jahren 1891-1893*; Berlin

Bere, Rennie M.   1955 *Exploration of the Ruwenzori*; Uganda Journal, *19*, 2: 121-136

 – 1966, *The Way to the Mountains of the Moon*; Arthur Barker

Bernard, Pierre   1959, *Montagnes de la Lune*; Hachette, Paris

Burton, Richard F.   1860 *The Lake Regions of Central Africa*; Longman, Green, Longman & Roberts

Busk, Douglas   1957, *The Fountains of the Sun*; Max Parrish

Claudius Ptolemæus   150 AD, *Geographia*, 1. 9 & 4, 8 (quoting Marinus of Tyre, *Periplus of the Erythrean Sea* 77 AD)

Casati, Gaetano   1891, *Ten Years in Equatoria and the return with Emin Pasha*; Frederick Warne

Cooke, Peter & Doornbos, Martin   1982, *Rwenzururu Protest Songs*, Africa, *52*, 1: 37-60

Crotty, Raymond   1986, *Ireland in Crisis: a Study in Capitalist Colonial Underdevelopment*; Brandon, Dingle, Eire

Doornbos, Martin R.   1970, *Kumanyana and Rwenzururu: Two responses to ethnic Inequality* in *Protest and Power in Black Africa*, (Eds Robert I. Rotberg & Ali M. Mazrui); Oxford University Press, New York

de Fillipi, Fillipo   1908, *Ruwenzori: an account of the expedition of H.R.H. Prince Luigi Amedeo of Savoy, Duke of the Abruzzi*; Archibald Constable

Fishlock C.W.L. and Hancock G.L.R.   1933, *Notes on the Flora and Fauna of Ruwenzori with special reference to the Bujuku Valley*, Journal of the East African & Uganda Natural History Society, 44, 205-229, Nairobi

Fossey, Dian   1983, *Gorillas in the Mist*, Houghton Mifflin, USA & Hodder & Stoughton, UK

Freshfield D.W.,   1906, *Ruwenzori*, Alpine Journal, *23*, 171: 45-50; 175: 310-313

 – 1907, *Towards Ruwenzori*, Alpine Journal, *23*, 172: 87-98, 173: 185-202

Gessi, Romolo   1892, *Seven Years in the Soudan*; Sampson Low

Goldsmith, Edward   1989, *Development: The Cause of the Population Explosion*, The Ecologist, *19*, 1: 2-3

von Götzen, Graf A.   1895, *Durch Afrika von Ost nach West*, Berlin

Grant, James Augustus   1864, *A Walk across Africa, or Domestic Scenes from my Nile Journal*; William Blackwood

de Grunne, X.   1937, *Le Ruwenzori*; Dupriez, Brussels

Hedberg, Olov   1951, *Vegetation Belts of the East African Mountains: Results of the Swedish East African Expedition 1948, Botany No 1*, Särtryck ur: Svensk Botanisk Tidskrift, Bd 45, H. l. Uppsala, Sweden

 – 1964, *Features of Afroalpine Plant Ecology*, Acta Phytogeographica Suecica, Edidit Svenska Växtageografiska Sällskapet 49, Uppsala, Sweden; Almqvist & Wiksells Boktryckeri

 – 1969, *Growth Rate of the East African Giant Senecios*, Nature 222, 5189: 163-164

 – 1969, *Evolution and speciation in a tropical high mountain flora*, Biol. J. Linn. Soc., 1: 135-148

 – 1970, *Evolution of the Afroalpine Flora*, Biotropica, *2*, 1: 16-23

 – 1973, *Adaptive Evolution in a Tropical-Alpine Environment*, in Taxonomy and Ecology (Ed. V.N. Heywood), The Systematics Ass. Spec. Vol. No. 5, London & New York

Herodotus   486-408 BC, Book 2, *Euterpe* 28

Hicks, P.H.   1947 *The Portal Peaks of Ruwenzori*, Geographical Journal, *108*, 210-220

Howard, Peter   1988, *Proposal for the Establishment of a new National Park in the Rwenzori mountains, Uganda*; World Wildlife Fund International & New York Zoological Society

Humphreys, G. Noel   1927, *Ruwenzori*, Alpine Journal, *39*, 234: 99-104

 – 1927, *New Routes on Rwenzori*, Geographical Journal, *69*, 516-531

 – 1933, *Ruwenzori: flights and further explorations*, Geographical Journal, *82*, 481-514

Ingham, Kenneth   1975, *The Kingdom of Toro in Uganda*, Studies in African History – 10; Methuen

Johnston, Harry H.   1902, *The Uganda Protectorate*; Hutchinson

 – 1903, *The Nile Quest: a record of the exploration of the Nile and its Basin*; Lawrence & Bullen

— Bibliography —

Joseph Sibalinghana Matte   (undated), *A Brief Political History of Bakonzo – the Inhabitants of the Rwenzori Mountains*, Unpub. cyclostyled paper, 73pp

Kandt, Richard   1904, *Caput Nili: eine empfindslame Reise 3" den Quellen des Nils*; Dietrich Reimer (Ernst Vohfen), Berlin

Kafsir, Nelson   1970, *Toro District: Society and Politics*, Mawazo, 2, 3: 39-53

 – 1976, *Seizing Half a Loaf: Isaya Mukirane and Self-recruitment for Secession*, in The Making of Politicians: Studies from Africa and Asia, (Ed. W. Morris-Jones); Athlone Press

 – 1976, *Ethnic Political Participation in Uganda: 3, Rwenzururu* in The Shrinking Political Arena; University of California Press, Berkeley & Los Angeles

Krapf, Johann Lewis   1860, *Travels, Researches and Missionary Labours During an Eighteen years' Residence in Eastern Africa*, Trubner (2nd edn. Frank Cass, 1968)

Loveridge J.P.   1968, *Plant ecological investigations in the Nyamugasani valley Ruwenzori mountains, Uganda*; Kirkia, 6, 153-168

Lugard F.D.   1893, *The Rise of our East African Empire*; William Blackwood

Mason A.M.   1878, *Report of a Reconnaissance of Lake Albert: made by order of His Excellency General Gordon Pasha, Governor-General of the Soudan*, Proc. Roy. Geogr. Soc. (old ser) 22, 3 : 225-229

Moore J.E.S.   1901, *To the Mountains of the Moon*; Hurst & Blackett

Mowat, Farley   1987, *Woman in the Mists*, (the edited diaries of Dian Fossey); Macdonald

Osmaston, Henry A.   1967, *The sequence of glaciations in the Ruwenzori and their correlation with glaciations of other mountains in east Africa and Ethiopia*, Palaentology of Africa, 2, 26-28

Osmaston H.A. & Pasteur, D.   1972, *Guide to the Ruwenzori: The Mountains of the Moon* (includes a bibliography of the range up to 1968); Mountain Club of Uganda & West Col Productions, 1 Meadow Close, Goring-on-Thames, Reading, Berkshire, U.K.

Ormerod W.E.   1980-81, *The African Husbandman and his Diseases: a moral dilemma posed by development versus ecological stability*, Rural Africana, 8-9, (Fall-Winter 1980-81) 133-147

Parc National des Volcans   1985, *Management Plan, Pts I & II (2nd draft)*, L'Office Rwandais du Tourisme et des Parcs Nationaux (ORTPN), Kigali, Rwanda

Parke, Thomas Heazle   1891, *My Personal Experiences in Equatorial Africa*; Sampson Low

Pasteur D.   1964, *An expedition to the Emin and Gessi area on Ruwenzori*, Alpine Journal, 69, 309: 191-200

Perham, Margery   1956, *Lugard: The Years of Adventure 1858-1898*; Collins

Schaller, George   1965, *The Year of the Gorilla*; Collins

Schlichter, Henry   1891, *Ptolemy's Topography of Eastern Equatorial Africa*, Procs. of the Royal Geographical Society, (New Ser.) 13, 513-553 & map 576

Scott Elliot G.F.   1896, *A Naturalist in mid-Africa: being an account of a journey to the Mountains of the Moon and Tanganyika*; A.D. Innes

Shipton, Eric E.   1932, *Mountains of the Moon*, Alpine Journal, 44, 244: 88-96

 – 1943, *Upon that Mountain*; Hodder & Stoughton

Speke, John Hanning   1863, *Journal of the Discovery of the Source of the Nile*; William Blackwood

 – 1864, *What led to the Discovery of the Source of the Nile*; William Blackwood

Stacey, Tom   1965, *Summons to Ruwenzori*; Secker & Warburg

Stanley, Henry Morton   1878, *Through the Dark Continent*; Sampson Low

 – 1890,   *In Darkest Africa or The Quest, Rescue and Retreat of Emin, Governor of Equatoria*; Sampson Low, Marston, Searle & Rivington

 – 1914, *Autobiography of Henry M. Stanley* (Ed. Dorothy Stanley); Sampson Low

Stulhmann, Franz   1894, *Mit Emin Pascha ins Herz von Afrika*; Reimer, Berlin

Sayahuka, Muhindo   1983, *The origin and development of the Rwenzururu movemnet: 1900-1962*; MAWAZO (Journal of the Arts and Social Sciences faculties of Makerere University, Uganda) 5, 2:60-74

Synge, Patrick M.   1937, *Mountains of the Moon*; Drummond (and 1985 reprint by Waterstone)

Tanzania: crisis & struggle for survival   1986, Eds. Jannik Boesen, Kjell K. Havenik, Juhani Koponen, Rie Odgaard Scandinavian Institute for African Studies, Uppsala, Sweden

Taylor B.K.   1962, *The Western Lacustrine Bantu: the Konjo*, International African Institute, London

Thiam, Awa   1978, *Black Sisters, Speak Out*; Pluto

Tilman H.W.   1937, *Snow on the Equator*; Bell

Uganda-Congo Boundary Commission   1910, *Agreement between Great Britain and Belgium settling the boundary between Great Britain and the Congo, Brussels May 14th 1910*, British & Foreign State Papers 107 (1914, Pt 1) 348-349

Uganda Government   1962, *Report of the Commission of Inquiry into the Recent Disturbances amongst the Baamba and Bakonjo People of Toro*, Government Printer, Entebbe, Uganda

Uganda Now: Between Decay and Development   1988, Eds. Hölger Bernt Hansen & Michael Twaddle, East African Studies; James Currie, London & Heinemann, Nairobi

Vienne, Gerard et Guy   1980, *Akagera: des Lions du Nil aux Gorilles des Monts de la Lune*; Flammarion, Paris

Wollaston A.F.R.   1908, *From Ruwenzori to the Congo*; Murray

Wood, Michael   1987, *Different Drums*; Century

Woosnam R.B.   1907, *Ruwenzori and its life Zones* (account of the British Museum 1906 Expedition), Geographical Journal, 30, 630-642

Yeoman, Guy H.   1985, *Can the Rwenzori be Saved?* SWARA, (East African Wildlife Journal) 8, 3: 8-12

MAPS
The definitive series is the 1:50,000 set of 10 sheets published by the Ugandan Lands & Surveys Department in 1958. The central massifs and lake regions are covered by sheet number 65/2 (Margherita) and the approach up the Mubuku valley by sheet 66/1 (Mubuku). The rest of the range in Uganda and its outlying regions are covered by sheets 56/1, 56/2, 56/3, 56/4, 66/2, 66/3, 66/4 and 65/4. There is also A 1:25,000 enlargement of the central peaks.

Osmaston & Pasteur's pocket *Guide to the Ruwenzori* (1972) contains useful route maps as end papers.

Andrew Wielochowski's *Map & Guide* (1989) provides clear maps as well as route descriptions and advice to visitors.
  These last two items are published by West Col Productions, Goring, Reading, Berks, U.K., RG8 9AA

INDEX

Front endpapers. *The uppermost of the chain of three Kachope lakes which, at 13,000 feet form the headwaters of the Butawu stream that flows westwards into the Semliki river in Zaire. The light-green trees are orange-flowered St John's Wort,* Hypericum bequaertii, *the darker trees are phillipia giant heaths. Our camp was exquisitely sited under the trees of the moraine on the left of the picture. Half hidden in the mist is the noble cirque of Okusoma. The only inhabitants were a pair of white-spotted black Rwenzori ducks. There are approximately 25 lakes in central Rwenzori and this one is perhaps my favourite.*